Current Communications
IN MOLECULAR BIOLOGY
Cold Spring Harbor Laboratory / 1988

Cell Cycle Control in Eukaryotes

Edited by
David Beach
Cold Spring Harbor Laboratory
Claudio Basilico
New York University Medical Center
John Newport
University of California, San Diego

Titles in
Current Communications in Molecular Biology

PLANT INFECTIOUS AGENTS
ENHANCERS AND EUKARYOTIC GENE EXPRESSION
PROTEIN TRANSPORT AND SECRETION
IMMUNE RECOGNITION OF PROTEIN ANTIGENS
EUKARYOTIC TRANSCRIPTION
PLANT CELL/CELL INTERACTIONS
TRANSLATIONAL CONTROL
COMPUTER GRAPHICS AND MOLECULAR MODELING
MICROBIAL ENERGY TRANSDUCTION
MECHANISMS OF YEAST RECOMBINATION
DNA PROBES
ANGIOGENESIS: Mechanisms and Pathobiology
GENE TRANSFER VECTORS FOR MAMMALIAN CELLS
NUCLEAR ONCOGENES
INOSITOL LIPIDS IN CELLULAR SIGNALING
GENETIC IMPROVEMENTS OF AGRICULTURALLY IMPORTANT CROPS: Progress and Issues
MOLECULAR GENETICS OF PARASITIC PROTOZOA
ANTISENSE RNA AND DNA
VIRAL VECTORS
CELL CYCLE CONTROL IN EUKARYOTES

CELL CYCLE CONTROL IN EUKARYOTES

International Standard Book Number 0-87969-317-7
Book design by Emily Harste
Printed in the United States of America

Cover: Mitotic spindles in fission yeast visualized with anti-tubulin antibodies. Nuclei are stained with DAPI. (Photo courtesy of Professor Mitsuhiro Yanagida, Kyoto University, Japan.)

Conference Participants

Renato Baserga. Dept. of Pathology, Temple University School of Medicine, Philadelphia, Pennsylvania

Claudio Basilico, Dept. of Pathology, New York University Medical Center, New York

David Beach, Cold Spring Harbor Laboratory, New York

Rodrigo Bravo, European Molecular Biology Laboratory, Heidelberg, Federal Republic of Germany

Breck Byers, Dept. of Genetics, University of Washington, Seattle

W. Zacheus Cande, Dept. of Botany, University of California, Berkeley

Dan Chelsky, E.I. du Pont de Nemours & Co., Wilmington, Delaware

Fred Cross, Fred Hutchinson Cancer Research Center, Seattle, Washington

Giulio Draetta, Cold Spring Harbor Laboratory, New York

William Dunphy, Dept. of Biology, University of California, San Diego, La Jolla

William Earnshaw, Dept. of Cell Biology and Anatomy, Johns Hopkins University School of Medicine, Baltimore, Maryland

B. Robert Franza, Jr., Cold Spring Harbor Laboratory, New York

David M. Glover, Dept. of Biochemistry, Imperial College of Science and Technology, London, England

Edward Harlow, Cold Spring Harbor Laboratory, New York

Leland H. Hartwell, Dept. of Genetics, University of Washington, Seattle

Nathaniel Heintz, Rockefeller University, New York, New York

Kevin Hennessy, Massachusetts Institute of Technology, Cambridge

David Hilborn, Eastman Kodak Research Laboratories, Rochester, New York

David E. Hill, Genetics Institute, Cambridge, Massachusetts

Tim Hunt, Dept. of Biochemistry, University of Cambridge, England

Jeremy S. Hyams, Dept. of Biology, University College, London, England

Stefan Jentsch, Dept. of Biology, Massachusetts Institute of Technology, Cambridge

Ron Laskey, Dept. of Zoology, CRC Molecular Embryology Group, Cambridge, England

Stanley Lin, Schering Corporation, Bloomfield, New Jersey

James L. Maller, Dept. of Pharmacology, University of Colorado School of Medicine, Denver

Kunihiro Matsumoto, DNAX Research Institute, Palo Alto, California

N. Ronald Morris, Dept. of Pharmacology, UMDNJ-Robert Wood Johnson Medical School, Piscataway, New Jersey

Daniel Nathans, Dept. of Molecular Biology and Genetics, Johns Hopkins University School of Medicine, Baltimore, Maryland
John Newport, Dept. of Biology, University of California, San Diego, La Jolla
Takeharu Nishimoto, Dept. of Molecular Biology, Kyushu University, Fukuoka, Japan
Patrick O'Farrell, Dept. of Biochemistry and Biophysics, University of California School of Medicine, San Francisco
Harvey L. Ozer, Center for the Study of Gene Structure and Function, Hunter College, CUNY, New York, New York
Steve Reed, Dept. of Molecular Biology, Research Institute of Scripps Clinic, La Jolla, California
Jim Roberts, Fred Hutchinson Cancer Research Center, Seattle, Washington
Tom Roberts, Dana Farber Cancer Institute, Boston, Massachusetts
Joan Ruderman, Dept. of Zoology, Duke University, Durham, North Carolina
Philip Sass, Lederle Laboratories, Pearl River, New York
Charles D. Stiles, Dana Farber Cancer Institute, Boston, Massachusetts
Bruce Stillman, Cold Spring Harbor Laboratory, New York
Dale Vandre, Dept. of Biological Sciences, Southern Methodist University, Dallas, Texas
Mitsuhiro Yanagida, Dept. of Biophysics, Kyoto University, Japan
Paul Young, Cold Spring Harbor Laboratory, New York
Edward Ziff, Dept. of Biochemistry, New York University Medical Center, New York

The meeting Cell Cycle Control in Eukaryotes was funded entirely by proceeds from the Laboratory's Corporate Sponsor Program, whose members provide core support for Cold Spring Harbor and Banbury meetings:

Abbot Laboratories
American Cyanamid Company
Amersham International plc
Applied Biosystems
Becton Dickinson and Company
Boehringer Mannheim GmbH
Bristol-Myers Company
Cetus Corporation
Ciba-Geigy Corporation
Diagnostic Products Corporation
E.I. du Pont de Nemours & Company
Eastman Kodak Company
Genentech, Inc.
Genetics Institute
Hoffmann-La Roche Inc.
Johnson & Johnson
Life Technologies, Inc.
Eli Lilly and Company
Millipore Corporation
Monsanto Company
Oncogene Science, Inc.
Pall Corporation
Pfizer Inc.
Pharmacia
Schering-Plough Corporation
Smith Kline & French Laboratories
Tambrands Inc.
The Upjohn Company
The Wellcome Research Laboratories,
Burroughs Wellcome Co.
Wyeth Laboratories

Contents

Preface

One of the fundamental problems in biology is to understand in molecular detail how cells reproduce themselves. This subject has received great attention both from those who study the primary events of the cell-division cycle itself and from those whose main interest lies in the mechanisms of oncogenesis. The field has tended to be subdivided by a slightly artificial barrier: "growth control" on one side and "cell-cycle control" on the other side. This series of short papers arose from a meeting held at the Cold Spring Harbor Banbury Conference Center in March 1988 that was specifically intended to ignore this division. Thus, the meeting brought together people whose interests ranged from mammalian growth factor signaling pathways to the simple biphasic division cycles of early frog embryos.

We thank Dr. James Watson for his support and encouragement. We are also grateful to Jan Witkowski and Bea Toliver of the Banbury Conference Center for their help in organizing this meeting, and to Katya Davey for her gracious hospitality at Robertson House. Sincere appreciation is extended to Nancy Ford (Managing Director of Publications), Dorothy Brown, Nadine Dumser, and Mary Cozza for their efforts in publishing this book.

D.B.
C.B.
J.N.

Introduction

D. Beach,[1] C. Basilico,[2] and J. Newport[3]

[1]Cold Spring Harbor Laboratory, Cold Spring Harbor
New York 11724

[2]New York University Medical Center, New York, New York 10016

[3]University of California, San Diego, La Jolla
California 92093

As human longevity continues to increase, a paradox has emerged. Although the most tractable diseases are each in turn being rendered harmless, those that were once relatively insignificant in numerical terms, but for which effective remedies have yet to be found, are assuming an ever greater importance. Cancer is one of these diseases, and the motivation to develop more effective strategies of antitumor therapy is therefore considerable.

With the exception of surgery, the most widely used cancer treatments (e.g., chemotherapy and radiotherapy) exploit relatively crude biological differences between proliferative and nondividing cells. Consequently, in addition to the replicating tumor cells, normally dividing cells, such as those of the hematopoietic or immune system, may be damaged severely by the therapeutic regime. This leaves patients with only a limited capacity to withstand many forms of treatment.

The present inability to adopt more direct approaches toward tumor therapy, i.e., those that take advantage of specific properties of the diseases tissue, touches upon our lack of understanding of one of the most fundamental properties of all living matter, namely, its ability to reproduce by cell division. Elaboration of the molecular pathways that regulate cell growth and division may be expected to lead to far more subtle strategies of clinical intervention.

This series of short papers presents the thinking of many of the most active researchers engaged in the study of the molecular basis of cell-cycle regulation in eukaryotic cells. A wide range of organisms, including diatoms, yeasts, molds, clams, fruit flies, and frogs, in addition to mammalian cells, are described. Each system has its own peculiar experimental merits: accessibility of the ascomycete fungi to the techniques of classic genetics, the opportunity to study the rapid cell cycles of early embryos in the absence of cell growth, or simply the excellent material that some species provide for cell biological and microscopic studies.

The approach of seeking the best experimental material for a particular problem, regardless of phylogenetic origin, has one drawback that has often attracted limited attention from those whose primary interest is in human disease, namely, that as long as each genetic or cell biological system remained insulated in a cocoon of scientific language and experimental material, there was relatively little scope for truly productive cross-fertilization.

This problem has been alleviated dramatically by technical advances, particularly in the field of molecular biology. As cell-cycle-control genes and their encoded proteins are identified, sequenced, and characterized biochemically in various organisms, the basic unity of all eukaryotes with respect to cell-cycle mechanisims becomes increasingly apparent. This series of papers emphasizes the merits of seeking the best experimental material to approach a particular problem and allowing any misgivings that might surround the phylogenetic status of a chosen experimental material to vanish in the light shed upon a fundamental process.

Growth Control

Understanding how the cell-division cycle is regulated requires a grasp of two closely related but separable problems: (1) How does the cell determine whether to multiply, as opposed to being quiescent, or even terminally differentiated? (2) Once actively proliferating, how is progress through the stages of the division cycle monitored and regulated by the cell?

A widely exploited approach toward the study of the transition from quiescence to cell division has been to activate nondividing tissue culture cells by exposing them to either crude serum or purified growth factors. It is striking that the overall rate of protein synthesis may increase only twofold or less as cells are induced to replicate DNA and enter mitosis following exposure to growth factors. Furthermore, it is estimated that the complexity of the total cellular mRNA pool increases by less than 1%. Consequently, there is merit in seeking genes that are specifically activated in response to growth factors, as some may reasonably be expected to be effectors of the mitogenic response.

Using the strategy of differential colony hybridization, a number of laboratories have pursued this approach. Stiles (Rollins et al.), Bravo, and Nathans et al. (all references refer to this volume) describe the isolation of approximately 80 mitogen-induced genes. It is intriguing how many of these genes have been identified previously, either as known oncogenes (e.g., *fos, myc, jun*) or as components of basic cellular constituents such as the cytoskeleton (e.g., actin, integrin, fibronectin, or tropomyosin).

The receptors of most growth factors are tyrosine kinases. One of the earliest known responses to mitogens, such as platelet-derived growth factor (PDGF), is a burst of tyrosine phosphorylation. To understand how this growth signal is transduced into a cellular response, it is necessary to identify the critical substrates of the growth factor receptor tyrosine kinases. Some progress in this direction is discussed by Morrison et al., who propose that phosphatidylinositol kinase exists in a complex with the PDGF receptor and is phosphorylated in response to activation of the receptor by growth factor. Hyperphosphorylation of the c-*raf* oncogene product, associated with an increase in its protein kinase activity, is also reported to occur following PDGF exposure.

Viral models have been used for many years to study the molecular basis of oncogenesis, and numerous oncogenes were first identified because they were carried on oncogenic viruses. The E1A gene of adenovirus has been the focus of particular interest, in part, because it can cooperate with *ras* in cellular transformation. It has been thought that E1A acts by regulating the expression of other genes, as either a positive or a negative component. Velcich et al. describe experiments indicating that deleted derivatives of E1A lacking the repressing (or negative) transcriptional regulatory sequences may still cooperate with *ras* in transforming rat embryo fibroblasts. A very promising clue as to how E1A may function is described by Harlow (Buchkovich et al.). E1A proteins have been shown to exist in a complex with several cellular proteins. Deletion studies indicate that the binding of a polypeptide of 105 kD may be particularly important in transformation. This 105-kD protein has now been shown to be the product of the Rb oncogene, the loss of which is associated with the occurrence of retinoblastoma.

Regulation of G_1/S

Full understanding of cell-cycle control requires that the pathways of mitogenic signaling be interconnected with those that regulate specific steps in the division cycle. Initiation of DNA synthesis is one of the most important of these steps. The transition from G_1 to S phase has been studied from a genetic angle because it has proved to be very difficult to identify critical regulatory elements using nongenetic biochemical approaches.

For a variety of biological reasons, classic mutagenesis and genetic analyses are limited in mammalian cells, but some mutants defective in cell-cycle progression have been isolated in tissue culture cell lines. More recently, some of the relevant genes have been isolated by rescue of the defective mutant phenotype. This approach has been taken in the laboratories of Basilico (Basilico et al.), Nishimoto (Nishimoto et al.) and Ozer (Dermody et al.). This work emphasizes some of the difficulties in both obtaining and analyzing mammalian cell-cycle mutants. Four genes have been identified so far, but their biochemical function is still unclear. Furthermore, the isolation of genes by DNA-mediated gene transfer is very time consuming. Basilico et al. describe an approach toward the rescue of genes that makes use of the cDNA Okyama-Berg expression library in an SV40 vector. This has the merit that biologically active plasmids containing the gene of interest can be recovered following a second round of transfection by fusing the transfectants with *cos* cells.

Yeast has provided the best material for genetic analysis of the eukaryotic cell cycle. The division cycle of *Saccharomyces cerevisiae* is regulated primarily at a step in G_1 that is commonly referred to as Start. Many genes involved in the traverse of Start have been identified, including the *CDC28* gene, which encodes a protein kinase and is homologous to the *cdc2*$^+$ gene of *Schizosaccharomyces pombe*, and a similar gene in vertebrates. Reed et al. describe isolation of genes that can suppress temperature-sensitive *CDC28* mutants on multicopy vectors. One of these is a homolog of the *suc1*$^+$ gene of fission yeast (see Draetta and Beach; Dunphy et al.), and another has some homology with cyclins (see Minshull et al.; Ruderman). A variety of other genes required for the initiation of DNA synthesis have been described in this budding yeast. One of these, the product of *CDC46* (Hennessy and Botstein) is translocated into the nucleus in a cell-cycle-dependent manner.

In the cell cycle of budding yeast, the growth state of the cell is monitored at Start, and initiation of DNA synthesis does not occur until a minimal cell size has been attained. A mutant that commits to the cell cycle at an abnormally small cell size is described by Cross. Cells carrying this mutation are both very small and also defective in response to the α-mating factor. This pheromone acts to arrest the division cycle at Start, thereby allowing cells of opposite mating type to undergo sexual fusion. Matsumoto et al. describe various other genes involved in the pheromone response and also in the cAMP pathway, which plays a central role in growth regulation in this yeast.

Events that immediately precede or are coincident with the initiation of DNA replication have been investigated in both yeast and vertebrate cells. These include the synthesis of nucleotide precursors, enzymes involved in DNA replication, and histones. Expression of the histone gene family is subject to cell-cycle regulation. Maximal synthesis occurs during S phase. The elements within the histone genes and the *trans*-acting factors that bind to these elements are described by Heintz. Two other S-phase proteins, thymidine kinase and proliferating cell nuclear antigen (PCNA, an auxiliary protein of polymerase δ), are discussed by Baserga et al. Antisense oligonucleotides prepared against the PCNA gene are shown to inhibit synthesis of the PCNA gene product and thus to inhibit DNA replication.

The study of mammalian DNA replication has been stimulated greatly by development of in vitro models that faithfully mimic replication in vivo. One such system makes use of SV40 as the replication template and requires T antigen for efficient DNA synthesis. This protocol has been developed and used by Stillman et al. to identify and purify a variety of cellular factors that play a role in either the initiation or elongation steps of synthesis. PCNA is shown to be essential for leading-strand, but not lagging-strand, synthesis. The same experimental system is employed by Roberts (Roberts and D'Urso) to distinguish factors not only that are essential for replication, but that may be present in higher amounts in S-phase rather than G_1 cells. Such factors might be specifically involved in the initiation of DNA synthesis.

Finally, the problem of why DNA replication occurs only once per cell-division cycle is addressed by Laskey et al., who find that permeabilization of the nuclear membrane may allow a replicated nucleus to reinitiate DNA synthesis without an intervening mitogenic event. It is suggested that dissolution of the nuclear membrane at mitosis might allow entry into the nucleus of a factor that is essential for the subsequent round of DNA synthesis.

Regulation of Mitosis

The segregation of cellular components between daughter cells at mitosis is a remarkably ordered process. During mitosis, there is a dynamic rearrangement of many of the most important components within the cell, including condensation of the chromosomes, breakdown of the nucleus, rearrangement of both the actin and microtubule cytoskeletons, and inhibition of both endo- and exocytosis. A clear picture of how mitosis is regulated is beginning to emerge as a result of recent observations.

Previous experiments demonstrating that inhibition of DNA replication prevented cells from entering mitosis led investigators to conclude that completion of DNA synthesis was an obligatory prerequisite of entering mitosis. However, it was unclear how DNA replication suppressed mitotic events at the molecular level. The isolation and characterization of temperature-sensitive mutants that override the feedback control of replication on mitosis have now made this interesting problem approachable (see Nishimoto et al.; Morris; Weinert and Hartwell). In all cases, these mutations appear to inactivate a protein that suppresses mitosis before DNA replication (or repair) is completed. Furthermore, studies using these mutations indicate that cells contain most, if not all, of the mRNA needed to enter mitosis before replication is complete. Thus, many of the proteins needed during mitosis might be present in an inactive form during S phase. This raises the interesting possibility that these mitotic proteins may play a role in modulating normal interphase processes and that amplification of their interphase function results in mitosis.

During the last few years, great progress has been made in identifying the proteins that regulate the initiation of mitosis. Much of our current knowledge about this problem is derived from studying the simple biphasic cell cycle present during early development. These studies have strongly implicated the "cyclin" proteins as major regulators of mitosis (see Minshull et al.; Ruderman). The proteins accumulate rapidly just prior to mitosis and are subsequently degraded just before the metaphase-anaphase transition of the cell cycle occurs. Furthermore, the cyclins have been shown to be required for the G_2- to M-phase cell-cycle transition during maturation of the *Xenopus* egg. The cyclins have been conserved in numerous divergent animals (clams, sea urchins, frogs, yeast, and flies), and this class of proteins now appears to play a primary role in initiating mitosis.

A second factor implicated in regulating the initiation of mitosis is maturation promoting factor (MPF) derived from *Xenopus* eggs (see Dunphy et al.; Maller et al.). Induction of mitosis by MPF can be assayed by microinjection into *Xenopus* oocytes or addition to a cell-free extract made from oocytes. Unlike cyclins, which appear to be synthesized and degraded during each cell cycle, MPF is present throughout the division cycle and appears to be activated just prior to mitosis and inactivated before the metaphase-anaphase transition. This activation-inactivation cycle is probably driven by a phosphorylation-dephosphorylation modification that may be carried out by protein kinase activities associated with the cyclin proteins.

MPF has been purified (see Maller et al.) and appears to be a complex of two proteins, 34 kD and 42 kD. The 34-kD protein is the homolog of the 34-kD protein kinase initially identified as the gene product of the *CDC28* gene of *S. cerevisiae* and subsequently found to be conserved in *S. pombe*, mammalian cells, and amphibians (see Draetta and Beach; Dunphy et al.). Like MPF, p34 from mammalian cells is present throughout the cell cycle. However, the kinase activity associated with this protein is cyclically activated and inactivated at mitosis and interphase, respectively. The increase in kinase activity associated with p34 at mitosis is likely to be critical in activating other proteins required during mitosis.

Many of the structural rearrangements associated with mitosis appear to be dependent on modification of preexisting proteins via phosphorylation. Such modifications appear to play a key role in modulating the activity of the centriole with respect to the number of microtubules that can polymerize from this structure (see Vandre et al.), as well as in controlling spindle elongation during anaphase (see Cande et al.). It is also clear that topoisomerase II plays a major structural role in chromosome condensation-decondensation during mitosis. This association-disassociation of topoisomerase II, either with itself or with other scaffold proteins, might be regulated via a phosphorylation-dephosphorylation cycle (see Earnshaw and Heck).

The primary event that causes mitosis to terminate has yet to be identified. However, a number of secondary events have now been found that should prove useful in identifying the molecular factors causing termination. Several proteins are now known to be degraded at the end of mitosis (see Ruderman; Earnshaw and Heck). In particular, there appears to be an extremely rapid and selective degradation of cyclin. Clearly, identification of the protease involved in cyclin proteolysis would be an important step toward understanding how a cell decides the enter anaphase. The identification of mutants in *S. pombe* that block sister chromatid separation at the metaphase-anaphase transition provides another excellent avenue for determining how this transition is controlled (see Hirano et al.). These mutants are supersensitive to caffeine, a known inhibitor of cAMP degradation. This observation, in conjunction with the measured decrease in cAMP that occurs at the metaphase-anaphase boundary, suggests that cAMP levels may be an important controlling element in terminating mitosis.

The isolation of mutants that allow centrioles to duplicate and separate in the absence of nuclear division (see Glover et al.) and allow initiation of DNA replication in the absence of spindle pole body duplication (see Byers et al.) indicates that many of the substrates modified during mitosis can act independently of each other. Such independent behavior suggests that many of the dynamic rearrangements that occur during mitosis are activated in parallel rather than sequentially. Thus, information obtained from one pathway may be useful in understanding how the process, as a whole, is regulated during mitosis.

Structure and Function of Platelet-derived Growth Factor-inducible Gene Products

B.J. Rollins, P. Oquendo, and C.D. Stiles

Dana Farber Cancer Institute, Boston, Massachusetts 02115

Replicative DNA synthesis and cell division require the transcription and translation of discrete unique sequence genes. The role of inducible genes in regulation of animal cell division has been especially well characterized in 3T3 cell cultures and in T-lymphocyte cultures. In both of these cell systems, the mitogenic response can be resolved into two stages. The first stage, which we have termed "competence," is regulated by a single mitogen. For 3T3 cells, this mitogen is platelet-derived growth factor (PDGF). For T cells, concanavalin A is a competence-inducing agent. The "progression" of competent 3T3 cells and T cells through G_1 into S phase is controlled by other growth factors that are not relevant to the present discussion.

Our laboratory (Cochran et al. 1983) and others (Linzer and Nathans 1983; Hirschorn et al. 1984; Edwards et al. 1985; Sukhatme et al. 1987) have isolated and characterized discrete unique sequence genes that are induced promptly and dramatically when fibroblast cells are exposed to PDGF. Some of the genes that are induced in PDGF-treated fibroblast cells are also induced when T cells are exposed to concanavalin A. For example, induction of the *c-myc* proto-oncogene is common to both cell types (Kelly et al. 1983). Other PDGF-inducible genes appear to be more lineage-specific. The PDGF-inducible JE and KC genes, for example, are not induced in concanavalin-A-treated T cells (K. Kelly, pers. comm.). On the contrary, the interleukin-2 (IL-2) and IL-2 receptor genes are lineage-specific for T cells and are not expressed in PDGF-treated fibroblasts (D. Hall and C.D. Stiles, unpubl.).

Those laboratories that have isolated PDGF-inducible genes from fibroblasts by differential colony hybridization find that between 0.1% and 0.3% of the genes expressed in fibroblast cells are induced by PDGF. By extrapolating from the number of genes expressed in fibroblast cell cultures (10,000–30,000), it can be calculated that the "competence gene family" contains between 10 and 100 genes in total. In this paper, we review data concerning the role of mitogen-inducible competence genes in regulation of animal cell growth.

Role of c-*myc* and c-*fos* in the Mitogenic Response

A broad body of data now supports the view that the c-*myc* and c-*fos* proto-oncogene products function as intracellular mediators of the mitogenic response in fibroblasts, T cells, and many other cell systems. Many laboratories have placed the c-*myc* gene under transcriptional control of viral promoters and transfected the chimeric gene into recipient cells. When this is done, the recipient cells escape totally or partially from the growth requirement for factors that stimulate *myc*-gene expression in the parental cells (Armelin et al. 1984; Keath et al. 1984; Balk et al. 1985; Rapp et al. 1985; Kelekar and Cole 1986; Sorrenino et al. 1986; Stern et al. 1986; Cory et al. 1987). The same result has been achieved by microinjection of recombinant c-*myc* protein into cultured cells (Kaczmarek et al. 1985). Conversely, suppression of c-*myc* gene expression with antisense deoxyoligonucleotides inhibits the mitogenic response of concanvalin-A-stimulated T cells (Heikkila et al. 1987). Other gene-transfer experiments show that c-*myc* promotes growth and antagonizes differentiation of erythropoietic cell lines (Copola and Cole 1986; Dmitrovsky et al. 1986; Lachman et al. 1986; Prochowinik and Kukowska 1986; Wickstrom et al. 1988).

A somewhat smaller database supports the role of c-*fos* as an intracellular mediator of the mitogenic response. Experiments with antisense c-*fos* vectors indicate that induction of c-*fos* is critical to the mitogenic response of BALB/c-3T3 cells (Holt et al. 1986; Nishikura and Murray 1987). However, c-*fos* expression may be required only during the initial round of cell division when cells exit the G_0 growth arrest state. Studies by Nishikura and Murray (1987) suggest that c-*fos* expression is not necessary when 3T3 cell cultures are in an exponential growth state and cells pass from mitosis directly into G_1 of the next cell-division cycle.

Function of the Original Competence Genes, JE and KC

The very first PDGF-inducible genes to be isolated and characterized as such were JE and KC (Cochran et al. 1983). Recent studies from our laboratory indicate that JE and KC do not function as intracellular mediators of the mitogenic response. These genes encode proteins that appear to be secreted from the PDGF-treated cell. The JE gene encodes a secretory glycoprotein (B.J. Rollins et al., in prep.). The coding sequences of JE are homologous to several cytokines, including IL-2, IL-6, M-CGF, and α-interferon. Because of these structural linkages to known cytokines and because PDGF also induces two authentic cytokines, M-CSF and IL-6, we speculate that the JE gene product itself is a cytokine. However, if JE is a cytokine, its target cell and biological activities remain to be discovered.

The PDGF-inducible KC gene encodes the newest member of a growing superfamily of inducible proteins with ancestral linkage to the platelet α-granule protein platelet factor 4 (PF4) and connective tissue activating peptide III (P. Oquendo and C.D. Stiles, unpubl.). The KCgene product is actually the murine homolog of a gene termed *gro*, which was isolated recently by Anisowicz et al. (1987). Other proteins with sequence and apparent structural homologies with the KC/*gro* gene product include IP-10 (see Luster et al. 1985), 9E3 (see Bedard et al. 1987; Sugano et al. 1987), and 3-10 (see Schmid and Weissman 1987). These are all small secretory proteins whose expression is induced by γ-interferon, v-*src*, and staphylococcal enterotoxin, respectively. All of these genes including KC are linked, in an unclear way, to problems in wound healing and inflammation. However, target cells and biological functions of these proteins are not well resolved.

SUMMARY

PDGF stimulates expression of between 10 and 100 discrete unique sequence genes in its target cells. Some of these genes encode proteins that remain within the cell nucleus. A broad body of data supports the view that PDGF-inducible nuclear proteins function as intracellular mediators of the mitogenic response. Other PDGF-inducible gene products are secreted from the cell. Some of these secreted proteins have structural features that suggest functional homologies with T-cell cytokines. We speculate that these PDGF-inducible secretory proteins play a systemic role in wound-healing responses. However, biological functions and specific target cells for most of these proteins remain to be discovered.

ACKNOWLEDGMENTS

Work from our laboratory described in this paper was supported by research grants from the National Institutes of Health and the American Cancer Society. We are grateful to Rhonda Walker for assistance in preparing the manuscript.

REFERENCES

Armelin, H.A., M.C.S. Armelin, K. Kelly, T. Stewart, P. Leder, B.H. Cochran, and C.D. Stiles. 1984. Functional role for c-*myc* in mitogenic response to platelet-derived growth factor. *Nature* **310:** 655.

Anisowicz, A., L. Bardwell, and R. Sager. 1987. Constitutive overexpression of a growth-regulated gene in transformed Chinese hamster and human cells. *Proc. Natl. Acad. Sci.* **84:** 7188.

Balk, S.D., T.M. Riley, H.S. Gunther, and A. Morisi. 1985. Heparin-treated v-*myc* transformed chicken heart mesenchymal cells assume a normal morphology but are hypersensitive to epidermal growth factor (EGF) and brain fibroblast growth factor (bFGF); cells transformed by the vHa-ras oncogene are refractory to EGF and bEGF but are hypersensitive to insulin-like growth factors. *Proc. Natl. Acad. Sci.* **82:** 5781.

Bedard, P.-A., D. Alcorta, D.L. Simmons, K.-C. Luk, and R.L. Erikson. 1987. Constitutive expression of a gene encoding a polypeptide homologous to biologically active human platelet protein in Rous sarcoma virus-transformed fibroblasts. *Proc. Natl. Acad. Sci.* **84:** 6715.

Cavalieri, F. and M. Goldfarb. 1987. Growth factor-deprived BALB/c 3T3 murine fibroblasts can enter the S phase after induction of c-*myc* gene expression. *Mol. Cell. Biol.* **7:** 3554.

Cochran, B.H., A.C. Reffel, and C.D. Stiles. 1983. Molecular cloning of gene sequences regulated by platelet-derived growth factor. *Cell* **33:** 939.

Coppola, J.A. and M.D. Cole. 1986. Constitutive c-*myc* oncogene expression blocks mouse erythroleukemia cell differentiation but not commitment. *Nature* **320:** 760.

Cory, S., O. Bernard, D. Bowtell, S. Schrader, and J.W. Schrader. 1987. Murine c-*myc* retroviruses alter the growth requirements of myeloid cell lines. *Oncogene Res.* **1:** 61.

Dmitrovsky, E., W.M. Kuehl, G.F. Hollis, I.R. Kirsch, T.P. Bender, and S. Segal. 1986. Expression of transfected human c-*myc* oncogene inhibits differentiation of a mouse erythroleukemia cell line. *Nature* **322:** 748.

Edwards, D.R., C.L.J. Parfett, and D.T. Denhardt. 1985. Transcriptional regulation of two serum-induced RNAs in mouse fibroblasts; equivalence of one species of B2 repetitive elements. *Mol. Cell. Biol.* **5:** 3280.

Heikkila, R., G. Schwab, E. Wickstrom, S.L. Loke, D.H. Pluznik, R. Watts, and L.M. Neckers. 1987. A c-*myc* antisense oligodeoxynucleotide inhibits entry into S phase but not progress from G_0 to G_1. *Nature* **328:** 445.

Hirschhorn, R.R., P. Aller, Z.-A. Yuan, C.W. Gibson, and R. Baserga. 1984. Cell-cycle specific cDNAs from mammalian cells temperature-sensitive for growth. *Proc. Natl. Acad. Sci.* **81:** 6004.

Holt, J.T., T. Venkat Gopal, A.D. Moulton, and A.W. Niehnuis. 1986. Inducible production of *c-fos* anti-sense RNA inhibits 3T3 cell proliferation. *Cell Biol.* **83:** 4794.

Kaczmarek, L., J.K. Hyland, R. Watt, M. Rosenberg, and R. Baserga. 1985. Microinjected c-*myc* as a competence factor. *Science* **228:** 1313.

Keath, E.J., P.G. Caimi, and M.D. Cole. 1984. Fibroblast lines expressing activated c-*myc* oncogenes are tumorigenic in nude mice and syngeneic animals. *Cell* **39:** 339.

Kelekar, A. and M.D. Cole. 1986. Tumorigenicity of fibroblast lines expressing the adenovirus e1a, cellular p53, or normal c-*myc* genes. *Mol. Cell. Biol.* **6:** 7.

Kelly, K., B.H. Cochran, C.D. Stiles, and P. Leder. 1983. Cell-specific regulation of the c-*myc* gene by lymphocyte mitogene and platelet-derived growth factor. *Cell* **35:** 603.

Lachman, H.M., G. Cheng, and A.I. Skoultchi. 1986. Transfection of mouse erythroleukemia cells with *myc* sequences changes the rate of induced commitment to differentiate. *Proc. Natl. Acad. Sci.* **80:** 4271.

Luster, A.D., J.C. Unkeless, and J.V. Ravetch. 1985. Gamma interferon transcriptionally regulates an early-response gene containing homology to platelet proteins. *Nature* **315:** 672.

Nishikura, K. and J.M. Murray. 1987. Antisense RNA of proto-oncogene c-*fos* blocks renewed growth of quiescent 3T3 cells. *Mol. Cell. Biol.* **7:** 639.

Prochownik, J. and J. Kukowska. 1986. Deregulated expression of c-*myc* by murine erythroleukemia cells prevents differentiation. *Nature* **322:** 848.

Rapp, U.R., J.L. Cleveland, K. Brightman, A. Scott, and J.N. Ihle. 1985. Abrogation of IL-3 and IL-2 dependence by recombinant murine retroviruses expressing v-*myc* oncogenes. *Nature* **317:** 434.

Schmid, J. and C. Weissman. 1987. Induction of mRNA for a serine protease and a beta-thromboglobulin-like protein in mitogene-stimulated human leukocytes. *J. Immunol.* **139:** 250.

Sorrentino, V., V. Drozdoff, M.D. McKinney, L. Zeitz, and E. Fleissner. 1986. Potentiation of growth factor activity by exogenous c-*myc* expression. *Proc. Natl. Acad. Sci.* **83:** 8167.

Stern, D.F., A.B. Roberts, N.S. Roche, M.B. Sporn, and R.A. Weinberg. 1986. Differential responsiveness of myc- and ras-transfected cells by epidermal growth factor. *Mol. Cell. Biol.* **6:** 870.

Sugano, S., M.Y. Stoeckle, and H. Hanafusa. 1987. Transformation by Rous sarcoma virus induces a novel gene with homology to a mitogenic platelet protein. *Cell* **49:** 321.

Sukhatme, V.P., S. Kartha, F.G. Toback, R. Taub, R.G. Hoover, and C.-W. Tsai-Morris. 1987. A novel early growth response gene rapidly induced by fibroblast, epithelial cell and lymphocyte mitogenes. *Oncogene Res.* **1:** 343.

Wickstrom, E.L., T.A. Bacon, A. Gonzalez, D.L. Freeman, G.H. Lyman, and E. Wickstrom. 1988. Human promyelocytic leukemia HL-60 cell proliferation and c-*myc* protein expression are inhibited by an antisense pentadecadeoxynucleotide targeted against c-*myc* mRNA. *Proc. Natl. Acad. Sci.* **85:** 1028.

Complexity of the Early Genomic Response to Growth Factors in Mouse Fibroblasts

R. Bravo

European Molecular Biology Laboratory, 6900 Heidelberg
Federal Republic of Germany

A detailed knowledge of the biochemical events and the identification of the genes that are involved in the response to growth factors is currently being accumulated. Growth-factor-regulated genes may play a critical role in cellular transformation. This is supported by the findings that c-*fos* and c-*myc* present an immediate change in expression following stimulation of fibroblasts (Kelly et al. 1983; Greenberg and Ziff 1984; Kruijer et al. 1984; Müller et al. 1984) and also emphasizes that growth-controlling genes operate during the G_0 to G_1 transition. Hence, the analysis of the early genetic response of a resting cell to a mitogenic stimulus could lead to the identification of new genes that play an important role in growth control or differentiation. According to estimations, at present, a small percentage of the total number of genes responsive to growth factors in fibroblasts have been identified (Cochran et al. 1983; Lau and Nathans 1987; Lim et al. 1987). Therefore, new attempts are required to complement the previous work. With this aim, a comprehensive study of the genes induced during the G_0 to G_1 transition was performed. By differential screening and cross-hybridization, more than 80 mRNAs have been demonstrated to be immediately induced by serum. Some of these genes are described below.

RESULTS

Searching for Serum-inducible Genes

A λcDNA library prepared from quiescent 3T3 cells stimulated for 4 hours with 20% serum in the presence of cycloheximide was screened for cDNAs corresponding to the specific mRNAs induced by serum. Approximately 2 x 10^5 plaques were screened, and 2540 phage were found to hybridize specifically against cDNA from stimulated cells (Almendral et al. 1988).

To determine the number of different mRNAs represented

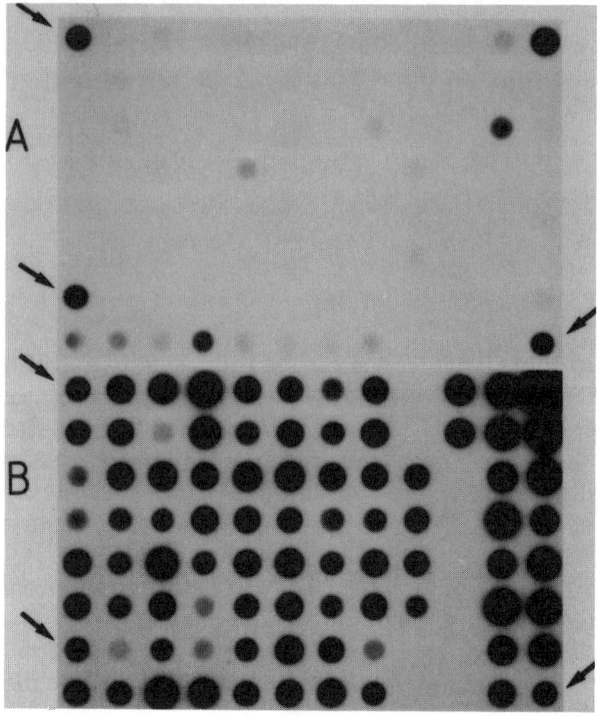

Figure 1 cDNA clones representing growth-factor-inducibile genes. Denatured recombinant pUC19 plasmid (1 μg) containing each insert was spotted onto GeneScreen Plus membrane. Filters were hybridized against high specific activity ^{32}P-labeled cDNA probes of cytoplasmic poly(A)$^{+}$ RNA from quiescent cells (*A*) or cells stimulated with 20% serum for 4 hr in the presence of cycloheximide. Plasmid containing glyceraldehyde 3-phosphate dehydrogenase (GAPDH) was used as control (→).

in the cDNA clones selected by differential screening, the inserts were tested for sequence homology by cross-hybridization. Those showing cross-hybridization under high-stringency conditions were considered as derived from an identical mRNA. After all phage were tested, 80 independent sequences were found to be present in the selected cDNAs and proved to be induced by serum in the presence of cycloheximide (Fig. 1).

To characterize the mRNAs from the selected cDNA clones, a detailed study of their kinetics of induction, half-life, and transcriptional activation was performed (Almendral et al. 1988). The results demonstrated that the mRNAs present two clear types of induction. A first group shows a time course of induction very similar to that described for c-*fos* mRNA, reaching its maximum level of expression approximately 30 minutes after stimulation and decaying rapidly to undetectable levels at 2 hours. The second group is represented by mRNAs whose maximum levels are reached after 2 hours of stimulation and slowly decrease thereafter or remain high for 8 hours or more. The half-life of some mRNAs is as short as 10–15 minutes, but others present a half-life of 4 hours or more.

Transcriptional analysis of these genes revealed that some present a very transient induction, reaching maximal levels at 15–30 minutes, and returning to basal levels within 60 minutes. In contrast, other genes presented an increased level of transcription for at least 8 hours.

Identification of the Genes Induced by Serum

Determining the sequence of the inducible mRNAs and their products will be essential for the understanding of the genetic program involved in the events leading to cell proliferation, as well as for studies concerning the possible role of these genes in cell differentiation. Therefore, we decided (1) to partially sequence all the selected cDNA clones and to determine their homology with genes described previously, and (2) to use specific probes from putative regulatory genes in order to study whether related sequences were present in our collection of cDNAs. The results obtained so far are summarized in Table 1.

Genes Encoding Nuclear Proteins

Sequence comparison and cross-hybridization studies have shown that clones P49 and V59 correspond to c-*fos* and c-*myc*, respectively. The sequence of the complete cDNA clone for AC16 (Chavrier et al. 1988) was found to encode three domains similar to those present in "zinc finger proteins," such as TFIIIA (Vincent 1986), Krüppel (Schuh et al. 1986), and Sp1 (Kadonaga et al. 1987). The corresponding gene has been named Krox-20 (Chavrier et al. 1988). Recent studies have demonstrated that this protein has a half-life of approximately 20–30 minutes (M. Zerial, unpubl.). Interestingly, another gene highly homologous to Krox-20 in the zinc finger domains (94%) has recently been shown to be induced in several systems (Milbrandt 1987; Lemaire et al. 1988; Sukhatme et al. 1988). This corresponds to clone M57 in our collection.

The complete sequence of cloned AC113 revealed that it encodes a protein related to c-*fos*. Comparison of the amino acid

Table 1 Identified Products Induced by Growth Factors

Clone	Coded protein
AC16	zinc finger
M57	zinc finger
P49	c-*fos*
V59	c-*myc*
AC113	c-*fos* related
AH119	c-*jun* A
B31	c-*jun* B
AF21	c-*rel*
N51	*gro*-related
TT13	helix destabilizing
A15	plasminogen activator inhibitor
C15	tissue factor related
B2	β-actin
V58	α-tropomyosin
P38	β-subunit fibronectin receptor
TT1	fibronectin
V101	putative actin binding

sequences demonstrates that AC113 codes for a protein that has 95% homology with c-*fos* protein in the putative DNA-binding domain region. Their kinetics of induction, half-lives of the mRNAs, and transcriptional activation are very similar (M. Zerial, in prep.).

Clone AH119 has been demonstrated to correspond to the cellular homolog of v-*jun* (Bohmann et al. 1987; Maki et al. 1987; Angel et al. 1988). Its complete amino acid sequence shows a 98% homology with AP-1 and human c-*jun* sequences (R.-P. Ryseck, in prep.). Northern blot analysis revealed two mRNAs of approximately 2.7 and 3.2 kb. Evidence indicates that these mRNAs are transcripts of the same gene but with a different length at their 3′ ends. Another related sequence, described by Ryder et al. (1988) (c-*jun* B), is also present among our cDNAs and corresponds to clone B31.

All of these genes show a very rapid and transient expression after stimulation. Their mRNAs present half-lives of 10–15 minutes, and in all cases, cycloheximide increases their stability and prolongs the transcription activation period.

Genes Encoding Cytoskeletal-Matrix Proteins

Four clones have been identified to encode components of the cytoskeleton or extracellular matrix. Clones B2, P38, TT1, and

V58 correspond to β-actin, β-subunit of fibronectin receptor, fibronectin, and α-tropomyosin, respectively.

All four mRNAs present a long half-life, and their transcriptional levels remain high for at least 8 hours. These properties are very different from those of other immediate early genes.

The observation that cytoskeletal and matrix proteins are induced by serum growth factors is of significant biological interest considering that a similar response must be triggered during a normal wound-healing process in the organism.

Genes Encoding for Secreted Proteins

Three putative secreted proteins have been identified by sequence comparison. Clone C15 codes for a protein similar to human tissue factor; however, it is different. Clone A15 corresponds to plasminogen activator inhibitor II.

The sequence of the complete cDNA corresponding to clone N51 has demonstrated to present a 91% homology with *gro*, a gene described recently to be expressed in transformed Chinese hamster cells (Anisowicz et al. 1987). This protein is related to another protein shown to be induced in cells transformed by Rous sarcoma virus (Bedard et al. 1987; Sugano et al. 1987).

DISCUSSION

The results presented here demonstrate that induction of proliferation in quiescent cells by growth factors is accompanied by the onset of a complex genetic program. Eighty genes have been found to be induced immediately following serum stimulation. Possibly only a subset of these genes will be directly involved in triggering cell proliferation. We propose that some of the genes whose expression is very transient may not be essential during the cell cycle, but for the G_0 to G_1 transition, as demonstrated for c-*fos* (Nishikura and Murray 1987). Other genes presenting a more lasting change in expression could belong to a set of genes essential for the G_1 progression to S phase, as is the case of c-*myc* (Heikkla et al. 1987).

Interestingly, the majority of the nuclear proteins encoded by the inducible genes represent putative *trans*-acting factors (Table 1). These will possibly participate in the *trans*-activation of genes required for the progression to G_1 and S phases.

The availability of this large collection of genes will allow us to identify sets of genes that are specifically induced by one or several growth factors. This knowledge will be of great value for the identification of the regulatory elements responsible for the different transcriptional activation observed with each growth factor.

Studies on the expression of these genes during development will be of great interest, as they could play an important role in cell differentiation.

REFERENCES

Almendral, J.M., D. Sommer, H. Macdonald-Bravo, J. Burckhardt, J. Perera, and R. Bravo. 1988. Complexity of the early genetic response to growth factors in mouse fibroblasts. *Mol. Cell. Biol.* **8:** 2140.

Angel, P., E.A. Allegretto, S.T. Okino, K. Hattori, W.J. Boyle, T. Hunter, and M. Karin. 1988. Oncogene *jun* encodes a sequence-specific *trans*-activator similar to AP-1. *Nature* **332:** 166.

Anisowicz, A., L. Bardewell, and R. Sager. 1987. Constitutive overexpression of a growth-regulated gene in transformed Chinese hamster and human cells. *Proc. Natl. Acad. Sci.* **84:** 7188.

Bedard, P.A., D. Alcorta, D.L. Simmons, K.C. Luk, and R.L. Erikson. 1987. Constitutive expression of a gene encoding a polypeptide homologous to biologically active human platelet protein in Rous sarcoma virus-transformed fibroblasts. *Proc. Natl. Acad. Sci.* **84:** 6715.

Bohmann, F., T.J. Bos, A. Admon, T. Nishimura, P.K. Vogt, and R. Tjian. 1987. Human proto-oncogene c-*jun* encodes a DNA binding protein with structural and functional properties of transcription factor AP-1. *Science* **238:** 1386.

Chavrier, P., M. Zerial, P. Lemaire, J. Almendral, R. Bravo, and P. Charnay. 1988. A gene encoding a protein with zinc fingers is activated during G_0/G_1 transition in cultured cells. *EMBO J.* **7:** 29.

Cochran, B.H., A.C. Riffel, and C.D. Stiles. 1983. Molecular cloning of gene sequences regulated by platelet-derived growth factor. *Cell* **33:** 939.

Greenberg, M.E. and E.B. Ziff. 1984. Stimulation of 3T3 cells induces transcription of the *c-fos* proto-oncogene. *Nature* **311:** 433.

Heikkla, R., G. Schwab, E. Wickstrom, S.L. Lobe, D.H. Pluznik, R. Watt, and L.M. Neckers. 1987. A c-*myc* antisense oligodeoxynucleotide inhibits entry into S phase but not progress from G_0 to G_1. *Nature* **328:** 445.

Kadonaga, J.T., K.R. Carner, F.R. Masierz, and R. Tjian. 1987. Isolation of cDNA encoding transcription factor Sp1 and functional analysis of the DNA binding domain. *Cell* **51:** 1079.

Kelly, K., B.H. Cochran, C.D. Stiles, and P. Leder. 1983. Cell-specific regulation of the c-*myc* gene by lymphocyte mitogens and platelet-derived growth factor. *Cell* **35:** 603.

Kruijer, W., J.A. Cooper, T. Hunter, and I.M. Verma. 1984. Platelet-derived growth factor induces rapid but transient expression of the c-*fos* gene and protein. *Nature* **312:** 711.

Lau, L.F. and D. Nathans. 1987. Expression of a set of growth-related immediate early genes in BALB/c 3T3 cells: Coordinate regulation with c-*fos* and c-*myc*. *Proc. Natl. Acad. Sci.* **84:** 1182.

Lemaire, P., O. Revelant, R. Bravo, and P. Charnay. 1988. Two genes encoding potential transcription factors with identical DNA-binding domains are activated by growth factors in cultured cells. *Proc. Natl. Acad. Sci.* (in press).

Lim, R.W., B.C. Varnum, and H.R. Herschman. 1987. Cloning of tetradecanoyl phorbol ester-induced primary response sequences and their expression in density-arrested Swiss 3T3 cells and a TPA non-proliferative variant. *Oncogene* **1:** 263.

Maki, Y., T.J. Bos, C. Davis, M. Starbuck, and P. Vogt. 1987. Avian sarcoma virus 17 carries the *jun* oncogene. *Proc. Natl. Acad. Sci.* **84:** 2848.

Milbrandt, J. 1987. A nerve growth factor-induced gene encodes a possible transcriptional regulatory factor. *Science* **238:** 797.

Müller, R., R. Bravo, J. Burckhardt, and T. Curran. 1984. Induction of c-*fos* gene and protein by growth factors precedes activation of c-*myc*. *Nature* **312:** 716.

Nishikura, K. and J.M. Murray. 1987. Antisense RNA of proto-oncogene c-*fos* blocks renewed growth of quiescent 3T3 cells. *Mol. Cell Biol.* **7:** 639.

Ryder, K., L.F. Lau, and D. Nathans. 1988. A gene activated by growth factors is related to the oncogene v-*jun*. *Proc. Natl. Acad. Sci.* **85:** 1487.

Schuh, R., W. Aicher, V. Gaul, S. Cote, A. Preiss, D. Maier, E. Seifert, V. Nauber, C. Schroder, R. Kemler, and H. Jäckle. 1986. A conserved family of nuclear proteins containing structural elements of the finger protein encoded by Krüppel, a *Drosophila* segmentation gene. *Cell* **47:** 1025.

Sugano, S., M.Y. Stoeckle, and H. Hanafusa. 1987. Transformation by Rous sarcoma virus induces a novel gene with homology to a mitogenic platelet protein. *Cell* **49:** 321.

Sukhatme, V.P., X. Cao, L.C. Chang, C.H. Tsai, D. Stamenkovich, P.C.P. Ferreira, D.R. Cohen, S.A. Edwards, T.B. Shows, T. Curran, M.M. Le Beau, and E.D. Adamson. 1988. A zinc finger-encoding gene coregulated with c-*fos* during growth and differentiation, and after cellular depolarization. *Cell* **53:** 37.

Vincent, A. 1986. TFIIIA and homologous genes. The finger proteins. *Nucleic Acids Res.* **14:** 4385.

The Genomic Response to Growth Factors

D. Nathans, B. Christy, S. Hartzell, Y. Nakabeppu, and K. Ryder

Howard Hughes Medical Institute Laboratory and Department of Molecular Biology and Genetics, Johns Hopkins University School of Medicine, Baltimore, Maryland 21205

Cell growth factors, like many extracellular ligands, act through receptor-mediated pathways to alter gene expression. In quiescent cells in culture, addition of growth factor (or serum) leads to sequential expression of sets of genes, suggesting activation of a growth-related genetic program. The first set of genes expressed, which we have called immediate early genes, are transcriptionally activated even in the presence of an inhibitor of protein synthesis and therefore are thought to be activated directly by second messengers resulting from the growth-factor-receptor interaction. Among the immediate early genes are c-*fos* (Greenberg and Ziff 1984) and c-*myc* (Kelly et al. 1983) and genes that have been identified by isolation of cDNAs derived from mRNAs present in growth-factor-stimulated or serum-stimulated cells but not in quiescent cells (Cochran et al. 1983; Lau and Nathans 1985, 1987). Here, we describe the characterization of some of the cDNAs corresponding to immediate early genes of BALB/c-3T3 mouse cells.

Isolation of Immediate Early cDNA Clones

Lau and Nathans (1985) have described the procedure for preparing and screening a cDNA library for identifying immediate early genes. An important step is the preparation of mRNA from cells stimulated with serum in the presence of cycloheximide. Because inhibition of protein synthesis leads to superinduction of immediate early genes by serum or growth factors (Cochran et al. 1983), cDNAs of interest are enriched in the library and in the probe used for screening. Ten different cDNA clones were isolated from this library that fit the criteria of being derived from immediate early mRNAs, namely, absence of or barely detectable mRNA in quiescent cells, rapid transcriptional activation of the corresponding genes after stimulation by serum or platelet-derived growth factor (PDGF) or fibroblast growth factor (FGF), and superinduction by serum in the presence of cycloheximide (Lau and Nathans 1987). In all cases, gene transcription was transient, but transcription was prolonged in the presence of cycloheximide, suggesting that this set of genes may be down-regulated by transcription factor(s) encoded by one or more immediate early genes.

Nucleotide Sequence of cDNAs

We have analyzed five of the ten cDNAs corresponding to immediate early genes. Two of the cDNAs encode proteins whose amino acid sequences are not homologous to known proteins. The other sequences are more informative.

cDNA clone 482. The cDNA clone 482 encodes a protein of 294 amino acids, whose sequence and hydrophobicity profile suggest that it is a transmembrane protein; i.e., it has a hydrophobic signal-like peptide at the amino terminus, followed successively by a long stretch with four potential *N*-glycosylation sites, a short hydrophobic region, four basic residues, and a short carboxy-terminal sequence. These properties of the 482 protein are consistent with an orientation with respect to the plasma membrane in which most of the protein (including potential glycosylation sites) is extracellular and the short carboxyl terminus is intracellular. Experiments with antiserum raised against a portion of the protein indicate that protein 482 is on the surface of proliferating 3T3 cells. Although protein 482 is not related to proteins in the data banks, it is related to a recently reported protein sequence, namely, that of human tissue factor (hTF) (Spicer et al. 1987). Excluding signal sequences and transmembrane segments, the amino acid sequences of 482 and hTF show approximately 60% identity. It is therefore likely that gene 482 encodes the mouse tissue factor. Because hTF binds specifically to blood-clotting factor VII and activates its proteolytic activity (Nemerson 1988), it is tempting to speculate that tissue factors of both mouse and man may activate extracellular proteases involved in cell growth or cell movement.

cDNA clone 268. The cDNA clone 268 encodes a protein of approximately 533 amino acids. (There are four possible in-frame initiation codons near the 5′ end of the cDNA.) Two striking features of the sequence are a short stretch near the amino terminus rich in glycine and serine and a second region near the carboxyl terminus that has three direct repeats of a typical zinc finger, DNA-binding sequence. Clone 268 is the mouse homolog of the nerve growth factor (NGF)-induced mRNA (NGF-1) described by Milbrandt (1987). A similar cDNA clone was also isolated by Sukhatme and his colleagues (Cao et al. 1988). The zinc finger sequence of 268 is nearly identical to that reported recently for Krox 20 (Chavrier et al. 1988), but the rest of the protein sequence is different. Clone 268 mRNA is present at high concentration in mouse brain, thymus, and lung tissue.

cDNA clone 465. The cDNA clone 465 encodes a protein of 344 amino acids (Ryder et al. 1988). This is related to the v-*jun* oncoprotein (Vogt et al. 1987) or AP-1 (Bohmann et al. 1987; Angel et al. 1988). We therefore call 465 *junB*. *junB* has two regions of homology with v-*jun*, one near the amino terminus and another at the carboxyl terminus. The latter corresponds to the DNA-binding domain of v-*jun*/AP-1. Using *junB* cDNA as a probe, we have isolated mouse genomic and cDNA clones for other members of the *jun* family. *junA* is the murine c-*jun*, based on a comparison with human c-*jun*. Like *junB*, *junA* has the properties of an immediate early gene in BALB/c-3T3 cells. A third *jun*-related cDNA clone (*junD*) isolated from the 3T3 cell library has been partially sequenced. Its mRNA is present in quiescent cells and is only moderately increased following serum stimulation. Using antiserum against the probable DNA-binding domain of *junB*, we have detected a marked increase in *jun*-related proteins in 3T3 cells following serum stimulation. At least one of these proteins is phosphorylated.

DISCUSSION

The discovery that c-*fos* and c-*myc* are rapidly activated by growth factors and other cell ligands led to the belief that *fos* and *myc* proteins may be the primary regulators of the subsequent genomic response to these ligands. The situation now appears to be more complex. In addition to *fos* and *myc*, several genes encoding known or probable transcription factors are activated as part of the primary genomic response to growth factors: *junA* (c-*jun*), *junB*, 268, and Krox 20. The products of these genes may be involved in the shutoff of the immediate early genes and in the activation of genes expressed later in the genomic response to growth factors. Materials are in hand to explore these possibilities and to investigate further the mechanisms of gene activation by growth factors.

REFERENCES

Angel, P., E.A. Allegretto, S.T. Okino, K. Hattori, W.J. Boyle, T. Hunter, and M. Karin. 1988. Oncogene *jun* encodes a sequence-specific *trans*-activator similar to AP-1. *Nature* **332:** 166.

Bohmann, D., T.J. Bos, A. Admon, T. Nishimura, P. Vogt, and R. Tjian. 1987. Human proto-oncogene c-*jun* encodes a DNA binding protein with structural and functional properties of transcription factory AP-1. *Science* **238:** 1386.

Cao, Y., L. Chang, C.-H. Tsai-Morris, and V.P. Sukhatme. 1988. A novel growth response gene induced with *fos*-like kinetics by fibroblast, epithelial cell, and lymphocyte mitogens encodes a polypeptide with three DNA binding "finger" motifs. *J. Cell. Biochem.* (Suppl.) *12A:* 175 (Abstr. C745).

Chavrier, P., M. Zerial, P. Lemaire, J. Almendral, R. Bravo, and P. Charnay. 1988. A gene encoding a protein with zinc fingers is activated during G_0/G_1 transition in cultured cells. *EMBO J.* **7:** 29.

Cochran, B.H., A.C. Reffel, and C.D. Stiles. 1983. Molecular cloning of gene sequences regulated by platelet-derived growth factor. *Cell* **33:** 939.

Greenberg, M.E. and E.B. Ziff. 1984. Stimulation of 3T3 cells induces transcription of the c-*fos* proto-oncogene. *Nature* **311:** 433.

Kelly, K., B.H. Cochran, C.D. Stiles, and P. Leder. 1983. Cell specific regulation of the c-*myc* gene by lymphocyte mitogens and platelet-derived growth factor. *Cell* **35:** 603.

Lau, L.F. and D. Nathans. 1985. Identification of a set of genes expressed during the G_0/G_1 transition in cultured mouse cells. *EMBO J.* **4:** 3145.

———. 1987. Expression of a set of growth-related immediate early genes in BALB/c 3T3 cells: Coordinate regulation with c-*fos* and c-*myc*. *Proc. Natl. Acad. Sci.* **84:** 1182.

Milbrandt, J. 1987. A nerve growth factor-induced gene encodes a possible transcriptional regulatory factor. *Science* **238:** 797.

Nemerson, Y. 1988. Tissue factor and hemostasis. *Blood* **71:** 1.

Ryder, K., L.F. Lau, and D. Nathans. 1988. A gene activated by growth factors is related to the oncogene v-*jun*. *Proc. Natl. Acad. Sci.* **85:** 1487.

Spicer, E.K., R. Horton, L. Bloem, R. Bach, K.G. Williams, A. Guha, J. Kraus, T.-C. Lin, Y. Nemerson, and W.H. Konigsberg. 1987. Isolation of cDNA clones coding for human tissue factor: Primary structure of the protein and cDNA. *Proc. Natl. Acad. Sci.* **84:** 5148.

Vogt, P.K., T.J. Bos, and R.F. Doolittle. 1987. Homology between the DNA-binding domain of the GCN4 regulatory protein of yeast and the carboxyl-terminal region of a protein coded for by the oncogene *jun*. *Proc. Natl. Acad. Sci.* **84:** 3316.

Tyrosine Phosphorylation in Signal Transduction

D. Morrison,[1] D. Kaplan,[1] H. Piwnica-Worms,[1] T. Keller,[1] H. Mamon,[1] B. Cohen,[3] U. Rapp,[2] B. Schaffhausen,[3] L. Cantley,[3] and T. Roberts[1]

[1]Dana Farber Cancer Institute, Harvard Medical School, and [2]National Cancer Institute, Frederick, Maryland
[3]Tufts University Medical School, Boston, Massachusetts 02115

Our recent work has focused on the elucidation of the mechanisms by which membrane-bound tyrosine kinases transmit signals to the cell. To identify targets of tyrosine kinases, we have utilized immunoblot analysis using antiphosphotyrosine antibodies to examine cells expressing oncogene-encoded tyrosine kinases and cells in which tyrosine kinase growth-factor receptors have been activated by their ligands. Each activated tyrosine kinase gives a distinctive "fingerprint" of tyrosine phosphorylated proteins (Morrison et al. 1988). However, the pattern of apparent substrates was similar in certain cases, i.e., v-*fms*- and v-*sis*-transformed cells, and fibroblasts in which the normal CSF-1 and platelet-derived growth factor (PDGF) receptors have been activated, suggesting a common set of targets for the two receptors. We are currently in the process of identifying these "substrate" molecules. (Notably, although most of the proteins observed to be phosphorylated on tyrosine are almost certainly substrates of the given receptor or oncogene product, some substrates may be phosphorylated by other kinases that are activated in response to the initial event.) One route to identifying these tyrosine phosphorylated molecules involves classic biochemical purification of the individual substrates, followed by peptide sequencing and cloning of the genes that encode them. This type of analysis is in progress but has not yet yielded significant results. Another method of analysis has already proven fruitful and is discussed here. In this case, we have examined immunoprecipitates made with antiphosphotyrosine sera to look for candidate enzymatic activities that might be present. The logic is that to transmit signals effectively, the receptor must inevitably phosphorylate and modify the activity of certain cellular enzymes, the identity of which may be inferred from what is known of the general mechanisms of signal transduction. We have studied in some detail two such activities: a phosphatidylinositol kinase, which we have tentatively identified as an 85-kD protein, and a serine-threonine kinase, which seems to be identical to the protein product of the proto-oncogene *c-raf*.

The 85-kD Protein, a Candidate Phosphatidylinositol Kinase

We originally identified this 85-kD protein in anti-middle-T antigen immunoprecipitates from polyomavirus middle-T-antigen-transformed cells (Whitman et al. 1985). We have reported several lines of evidence strongly indicating that a phosphatidylinositol kinase activity is associated with this protein. In addition, mutant analysis demonstrates that the presence of the phosphatidylinositol kinase activity and the phosphorylation of the 85-kD protein in the middle T antigen immunoprecipitates correlates with the ability of middle T antigen to transform cells. Purification experiments indicate that the 85-kD protein tightly associates with a number of activated tyrosine kinases such as the PDGF receptor, $pp60^{v\text{-}src}$, the polyoma middle T/$pp60^{c\text{-}src}$ complex, and v-*fms*. Our current working hypothesis is that the activated tyrosine kinases mediate the phosphorylation of the 85-kD protein on tyrosine residues and thereby activate the phosphatidylinositol kinase activity. Consistent with this hypothesis, the 85-kD protein and phosphatidylinositol kinase activity are found in anti-phosphotyrosine immunoprecipitates from cells containing activated tyrosine kinases (Kaplan et al. 1987). Phosphatidylinositol kinase activity is enhanced 100-fold in antiphosphotyrosine immunoprecipitates from cells treated with PDGF and 10-fold in v-*fms*-transformed cells. The lipid kinase activity is unique in that it produces a 1,3-PIP, rather than the classic 1,4-PIP, thereby generating a new phosphatidylinositol metabolite that may perform a novel second messenger function in the phosphatidylinositol turnover pathway (M. Whitman et al., in prep.). Preliminary cell fractionation experiments indicate that the 85-kD protein is located primarily in membrane fractions.

As expected, when the 85-kD protein was immuno-precipitated from PDGF-treated cells labeled in vivo with $[^{32}P]$orthophosphate, the protein was found to be phosphorylated on both tyrosine and serine residues. Surprisingly, when the 85-kD protein is labeled in vitro by incubating antiphosphotyrosine immunoprecipitates with Mg^{++} and [γ-^{32}P]ATP, the protein is phosphorylated in roughly equivalent amounts on both tyrosine and serine residues, indicating that a serine kinase activity is also present in the antiphosphotyrosine immunoprecipitates.

The c-*raf* Protein is Phosphorylated and Activated in Response to Membrane Signals

A second substrate for a variety of activated tyrosine kinases was identified as the c-*raf* protein, the cellular counterpart of the known serine/threonine kinase v-*raf* (Moelling et al. 1984). In our studies, we have examined the phosphorylation and protein kinase activity of the c-*raf* protein in response to oncogenic transformation or mitogen treatment of mouse 3T3 cells. We have found that expression of a number of membrane-bound oncogene products, including v-*src*, v-*fms*, v-*ras*, and polyoma middle T antigen, caused hyperphosphorylation of the c-*raf* protein, resulting in a marked increase in its apparent size, whereas expression of nuclear oncogene products did not. Similar changes in c-*raf* (but not A-*raf*) protein also occurred rapidly in response to treatment of cells with PDGF, acidic fibroblast growth factor (FGF), epidermal growth factor (EGF), and phorbol-myristate-acetate (PMA), but not insulin. The phosphorylation occurred primarily on serine and threonine residues; however, we were able to detect a subpopulation of c-*raf* molecules that was phosphorylated on tyrosine residues in cells transformed by v-*src* and v-*fms* or in cells stimulated with PDGF. There may be several ways to achieve this phosphorylation because down-regulation of protein kinase C by chronic treatment with PMA abolishes c-*raf* activation by subsequent challenge with PMA but not by treatment with PDGF. Most importantly, the changes in phosphorylation that we observed are accompanied by the activation of c-*raf* kinase activity.

These experiments imply that the c-*raf* protein is involved in the mediation of proliferative signals generated at the membrane, but it is not involved in the mediation of proliferative signals originating in the nucleus. These results are supported by the experiments of Smith et al. (1986) and Rapp et al. (1988), suggesting that cytoplasmic oncogene products with associated serine/threonine kinase activity, such as v-*raf* and v-*mos*, function downstream from membrane-associated onco-

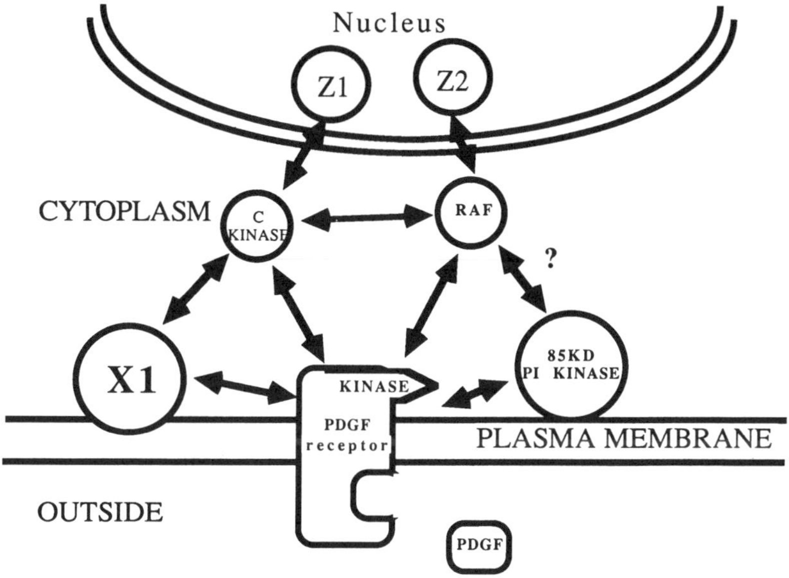

Figure 1 A schematic diagram of the roles of kinases and their substrates in signal transduction.

gene products in the signal transduction pathway. In the studies by Smith et al., microinjected antibodies directed against c-*ras* were able to inhibit the growth of cells transformed by v-*ras* and v-*src* but not those transformed by v-*raf* and v-*mos*. Rapp et al. observed that flat revertants of cells transformed by Kirsten murine sarcoma virus were able to suppress transformation by v-*ras* and v-*src* but not v-*raf*-carrying viruses. The investigators concluded that either the v-*raf* protein acts independently of $p21^{c\text{-}ras}$ or it functions downstream from $p21^{c\text{-}ras}$. The experiments presented here, showing c-*raf* protein phosphorylation in response to activated c-H-*ras* protein expression, suggest the latter possibility. We propose that the activation of c-*raf* by mitogenic stimuli at the membrane provides a new route of signal transduction to the nucleus. This route would function in addition to the well-established pathways mediated by protein kinases C and A and the calcium-regulated kinases. The need for alternative mechanisms for signal transduction is suggested by the observation that a number of serum-inducible promoters cannot be activated by any combination of phorbol esters, cAMP analogs, and calcium ionophores (Hall and Stiles 1987).

SUMMARY

We suggest that the activation of c-*raf* and of the 85-kD protein/phosphatidylinositol kinase by tyrosine kinases may serve to transduce signals from the membrane to cellular targets mediating proliferation. Although the 85-kD protein may function solely in the membrane, c-*raf* seems ideally poised to transfer signals directly to the cytoplasm and beyond. A schematic of our working model is presented in Figure 1.

REFERENCES

Hall, D.J. and C.D. Stiles. 1987. Platelet-derived growth factor inducible genes respond differentially to at least two distinct intracellular second messengers. *J. Biol. Chem.* **262:** 15302.

Kaplan, D.R., M. Whitman, B. Schaffhausen, D.C. Pallas, M. White, L. Cantley, and T.M. Roberts. 1987. Common elements in growth factor stimulation and oncogenic transformation: 85 kD Phosphoprotein and phosphatidylinositol kinase activity. *Cell* **50:** 1021.

Moelling, K., B. Heimann, P. Beimling, U.R. Rapp, and T. Sander. 1984. Serine-and threonine-specific protein kinase activities of purified *gag-mil* and *gag-raf* proteins. *Nature* **312:** 558.

Morrison, D.K., P.J. Browning, M.F. White, and T.M. Roberts. 1988. Tyrosine phosphorylates in vivo associated with v-*fms* transformation. *Mol. Cell. Biol.* **8:** 176.

Rapp, U.R., J.L. Cleveland, T.I. Bonner, and S.M. Storm. 1988. The *raf* oncogene. In *The oncogene handbook* (ed. P. Reddy et al.). Elsevier, Amsterdam. (In press.)

Smith, M.R., S.J. DeGudicbus, and D.W. Stacey. 1986. Requirement for c-*ras* proteins during viral oncogene transformation. *Nature* **320:** 540.

Whitman, M., D.R. Kaplan, B. Schaffhausen, L. Cantley, and T.M. Roberts. 1985. Association of phosphatidylinositol kinase activity with polyoma middle T antigen competent for transformation. *Nature* **315:** 239.

The Retinoblastoma Gene Product Binds the Adenovirus E1A Proteins

K.J. Buchkovich,[1] P. Whyte,[1,3] J.M. Horowitz,[2] S.H. Friend,[2] M. Raybuck,[1,4] R.A. Weinberg,[2] and E. Harlow[1]

[1]Cold Spring Harbor Laboratory, Cold Spring Harbor
New York 11724

[2]Whitehead Institute for Biomedical Research
Cambridge, Massachusetts 02142, and
Department of Biology, Massachusetts Institute of Technology
Cambridge, Massachusetts 02139

Human adenoviruses infect and proliferate in epithelial cells that line the upper respiratory tract and the intestines. The first virus-specific proteins that are synthesized following infection of these quiescent cells are the products of the early region 1A (E1A) (Nevins et al. 1979; Lewis and Mathews 1980). These proteins are multifunctional and carry out several roles essential for virus growth. The biochemical activities of the E1A proteins are not fully understood at present; however, these proteins appear to fulfill two general functions. First, the proteins are required for the transcriptional activation of the other viral early genes (Berk et al. 1979; Jones and Shenk 1979a; Nevins 1981; Ricciardi et al. 1981). Second, the E1A proteins stimulate the quiescent host cell to enter the cell cycle (Braithwaite et al. 1983; Spindle et al. 1985; Stabel et al. 1985; Kaczmarek et al. 1986; Zerler et al. 1987), presumably to induce the cellular machinery necessary for DNA synthesis. In nonpermissive cells that do not support viral DNA synthesis, the introduction of the E1A proteins either by adenovirus infection or by transfection of the cloned E1A region causes immortalization of primary cells (Houweling et al. 1980; Ruley 1983). Additional activities attributed to the E1A proteins are the ability to repress transcription from genes linked to certain viral or cellular enhancers (Borrelli et al. 1984; Velcich and Ziff

Present addresses: [3]Fred Hutchinson Cancer Center, Seattle, Washington 98104; [4]Amersham International plc, Forest Farm Industrial Estate, Whitchurch, Cardiff, Wales.

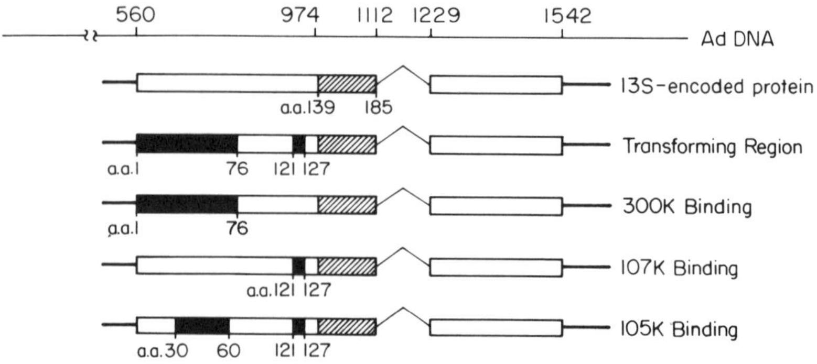

Figure 1 Comparison of the regions of E1A required for transformation and for the binding of the cellular proteins.

1985) and the ability to transform rodent cells in cooperation with a second oncogene, such as the *ras* oncogene or the adenovirus E1B gene (van der Eb et al. 1977; Graham et al. 1978; Jones and Shenk 1979b; Solnick and Anderson 1982; Chinnadurai 1983; Ruley 1983).

We and other investigators have demonstrated previously that the E1A proteins of adenovirus form stable complexes with several cellular proteins (Yee and Branton 1985; Harlow et al. 1986). The formation of these complexes may be a means by which the virus induces changes in normal cellular control. By identifying these cellular proteins and their involvement in the various E1A-mediated activities, we hope to gain insight into the biochemical mechanisms employed by E1A within the cell.

As an initial strategy to determine the possible functions of these complexes, we compared the regions of E1A required for binding to the cellular proteins with the regions necessary for various E1A activities. Mutational analyses have linked the binding of three of the cellular proteins, with approximate molecular weights of 105,000, 107,000, and 300,000, to the E1A-mediated activity of transformation. A series of E1A deletion mutants were constructed and screened for their ability to induce foci in primary baby rat kidney cells when cotransfected with an activated Ha-*ras* gene. Any mutation that resulted in a deletion within one of two regions of the E1A proteins, amino acids 1–76 or 121–127, destroyed focus-forming activity. Hence, these mutations have defined the minimal regions of the E1A proteins essential for transformation (Fig. 1).

The mutant E1A proteins were next screened for their ability to bind the cellular proteins. By infecting HeLa cells with recombinant adenoviruses containing the deletion mutations and then immunoprecipitating the E1A products and any bound cellular proteins, we have been able to map the binding sites of the 105K, 107K, and 300K proteins. The binding sites of the 300K and 107K proteins are amino acids 1–76 and 121–127, respectively, whereas the 105K protein requires the regions between amino acids 30–60 and 121–127 (Fig. 1). Interestingly, the boundaries of these binding sites coincided precisely with the boundaries of the transforming regions. Every E1A mutant that failed to bind one of these proteins also failed to transform cells efficiently, and, conversely, every mutant capable of binding all three of these proteins was transformation-competent. This result led us to hypothesize that these three proteins are cellular targets for E1A-mediated transformation (P. Whyte et al., in prep.).

The 105K E1A-associated protein (now renamed p105) took on increasing interest when it became apparent that it shared several characteristics with the polypeptide product of the retinoblastoma (Rb) gene (Lee et al. 1987), a gene first identified as a locus responsible for the predisposition to retinal tumors. In addition to having similar molecular weights, both the Rb polypeptide and the p105 protein are nuclear phosphoproteins that are present in most primate and rodent cells but are specifically absent from Rb tumor cell lines (Fig. 2).

To determine whether these two proteins were identical, we first confirmed that their molecular weights were the same, at least within the accuracy of SDS-PAGE. Immunoprecipitations were performed using either a monoclonal antibody specific for the E1A-bound p105 protein or a polyclonal antisera raised against a synthetic peptide that was designed according to the Rb cDNA sequence. The 105K proteins precipitated by these antibodies comigrated on polyacrylamide gels. We then compared these immunoprecipitated proteins by subjecting them to partial proteolysis with either *Staphylococcus aureus* V8 protease or *N*-chlorosuccinimide (NCS). V8 protease cleaves proteins enzymatically after acidic residues (aspartic acid or glutamic acid), whereas NCS cleaves chemically after tryptophan residues. Each reagent yielded an identical proteolysis pattern for both proteins, demonstrating that the positions of accessible aspartic acid, glutamic acid, and tryptophan residues are identical in the two polypeptides.

Finally, we performed two types of experiments to prove that these proteins were recognized directly by the anti-p105

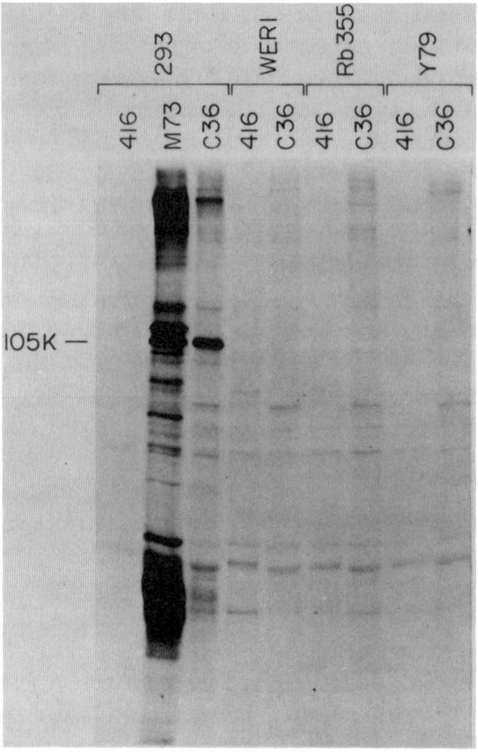

Figure 2 p105 is not present in Rb cell lines. 293 cells, human embryonic kidney cells transformed by the E1A and E1B regions of adenovirus, and three lines of Rb cells were radiolabeled with [^{35}S]methionine. The cells were lysed, and immunoprecipitations were performed as described by Harlow et al. (1986). PAb 416, C36, and M73 are mouse monoclonal antibodies specific for the SV40 large T antigen, the adenovirus E1A proteins, and the E1A-associated p105 protein, respectively.

and anti-Rb antibodies and were not simply spurious cross-reacting proteins or proteins immunoprecipitated via another antigen. In one set of experiments, the monoclonal antibody specific for p105 was shown to immunoprecipitate polypeptides produced by in vitro translations of cRNA transcribed directly from the cloned Rb cDNA. In complementary immunoblotting experiments, nitrocellulose-bound p105 was recognized by two antipeptide antibodies raised against different regions of the Rb proteins. By proving that p105 is recognized directly by more than one anti-Rb antibody, this last experiment demonstrated that p105 shares at least two epitopes with the Rb proteins. These data demonstrate that the E1A-associated p105 and the Rb proteins are products of the same gene.

The disruption of both copies of the Rb gene has been linked to the appearance of retinoblastomas and other genetically related tumors, suggesting that the Rb gene product may be a component of a regulatory pathway responsible for inhibiting cell proliferation. Because the inactivation of genes such as Rb leads to tumor growth, these proteins have been described as tumor suppressors or anti-oncogenes. The demonstration of the E1A/p105-Rb complex is the first example of an association between an oncogene and an anti-oncogene. The existence of this oncogene/anti-oncogene complex unifies two fields of research that were thought to be unrelated.

What is the role of the complex? We favor a model in which anti-oncogene proteins like p105 are elements of a pathway that allows cells to respond to environmental signals. These signals might be differentiation inducers, cell-cycle regulators, or other factors that carry an inhibitory cue to the cell. The loss of the Rb gene and subsequent loss of the Rb protein would remove or block the signal. As a consequence, the cell would lose its ability to respond normally to the inhibitory signal while retaining its proliferative ability. In this model, by interacting with p105, the E1A proteins may block the passage of the inhibitory signal and serve as an indirect stimulator of cell proliferation.

From the standpoint of the virus, Rb may be an attractive target for interfering with normal cell-cycling controls. One of the major goals of early viral proteins must be to drive quiescent cells into S phase. Therefore, the virus may be acting via the E1A/p105-Rb complex to produce an environment suitable for viral DNA synthesis.

This raises the questions of whether the other viral oncogenes also capitalize on the regulatory control of the Rb proteins and of whether they function by mimicking the normal cellular protein(s) that regulates the activity of the Rb gene product. Our hope is that future studies will reveal the mechanism by which Rb controls cell proliferation and the mechanism by which the cell or an invading virus may control Rb.

REFERENCES

Berk, A.J., F. Lee, T. Harrison, J. Williams, and P.A. Sharp. 1979. Pre-early adenovirus 5 gene product regulates synthesis of early viral messenger RNAs. *Cell* **17:** 935.

Borrelli, E., R. Hen, and P. Chambon. 1984. Adenovirus-2 E1A products repress enhancer-induced stimulation of transcription. *Nature* **312:** 608.

Braithwaite, A.W., B.F. Cheethan, P. Li, C.R. Parish, L.K. Waldron-Stevens, and A.J.D. Bellett. 1983. Adenovirus-induced alterations of the cell growth cycle: A requirement for expression of E1A but not E1B. *J. Virol.* **45:** 192.

Chinnadurai, G. 1983. Adenovirus *2/lp*⁺ locus codes for a 19 kd tumor antigen that plays an essential role in cell transformation. *Cell* **33:** 759.

Graham, F., T. Harrison, and J. Williams. 1978. Defective transforming capacity of adenovirus type 5 host range mutants. *Virology* **86:** 10.

Harlow, E., P. Whyte, B.R. Franza, Jr., and C. Schley. 1986. Association of adenovirus early-region 1A proteins with cellular polypeptides. *Mol. Cell. Biol.* **6:** 1579.

Houweling, A., P.J. Van den Elsen, and A.J. van der Eb. 1980. Partial transformation of primary rat cells by the leftmost 4.5% fragment of adenovirus 5 DNA. *Virology* **105:** 537.

Jones, N. and T. Shenk. 1979a. An adenovirus type 5 early gene function regulates expression of other early viral genes. *Proc. Natl. Acad. Sci.* **76:** 3665.

———. 1979b. Isolation of adenovirus type 5 host range deletion mutants defective for transformation in rat embryo cells. *Cell* **17:** 683.

Kaczmarek, L., B. Ferguson, M. Rosenberg, and R. Baserga. 1986. Induction of cellular DNA synthesis by purified adenovirus E1A proteins. *Virology* **152:** 1.

Lee, W.-H., J.-Y. Shew, F.D. Hong, T.W. Sery, and L.A. Donoso. 1987. The retinoblastoma susceptibility gene encodes a nuclear phosphoprotein associated with DNA binding activity. *Nature* **329:** 642.

Lewis, J.B. and M.B. Mathews. 1980. Control of adenovirus early gene expression: A class of immediate early products. *Cell* **21:** 303.

Nevins, J.R. 1981. Mechanism of activation of early viral transcription by the adenovirus E1A gene product. *Cell* **26:** 213.

Nevins, J.R., H.S. Ginsberg, J. Blanchard, M.C. Wilson, and J.E. Darnell, Jr. 1979. Regulation of the primary expression of early adenovirus transcription units. *J. Virol.* **32:** 727.

Ricciardi, R.P., R.L. Jones, C.L. Cepko, P.A. Sharp, and B.E. Roberts. 1981. Expression of early adenovirus genes requires a viral encoded acidic polypeptide. *Proc. Natl. Acad. Sci.* **78:** 6121.

Ruley, H.E. 1983. Adenovirus early region 1A enables viral and cellular transforming genes to transform primary cells in culture. *Nature* **304:** 602.

Solnick, D. and M.A. Anderson. 1982. Transformation-deficient adenovirus mutant defective in expression of region 1A but not region 1B. *J. Virol.* **42:** 106.

Spindler, K.R., C.Y. Eng, and A.J. Berk. 1985. An adenovirus early region 1A protein is required for maximal viral DNA replication in growth-arrested human cells. *J. Virol.* **53:** 742.

Stabel, S., P. Argos, and L. Philipson. 1985. The release of growth arrest by microinjection of adenovirus E1A DNA. *EMBO J.* **4:** 2329.

van der Eb, A.J., C. Mulder, F.L. Graham, and A. Houweling. 1977. Transformation with specific fragments of adenovirus DNAs. I. Isolation of specific fragments with transforming activity of adenovirus 2 and 5 DNA. *Gene* **2:** 115.

Velcich, A. and E. Ziff. 1985. Adenovirus E1a proteins repress transcription from the SV40 early promoter. *Cell* **40:** 705.

Yee, S.-P. and P.E. Branton. 1985. Detection of cellular proteins associated with human adenovirus type 5 early region 1A polypeptides. *Virology* **147:** 142.

Zerler, B., R.J. Roberts, M.B. Mathews, and E. Moran. 1987. Different functional domains of the adenovirus E1A gene are involved in regulation of host cell cycle products. *Mol. Cell. Biol.* **7:** 821.

Cell Growth Regulation by the Adenovirus-5 E1A Proteins

A. Velcich, D. Smith, and E. Ziff

Department of Biochemistry and Kaplan Cancer Center
New York University Medical Center
New York, New York 10016

DNA viruses that replicate in the nuclei of infected cells have evolved sophisticated mechanisms for utilizing cellular programs for the benefit of the viral life cycle. Among these are viral mechanisms, which stimulate cellular proliferative functions such as DNA synthesis. Cellular proliferative functions may be required by these viruses during the late stages of their life cycle, e.g., during replication of the viral genome. In fact, the papovaviruses and adenoviruses induce cellular DNA replication transiently during normal infections of permissive cells. When the viruses infect nonpermissive or semi-permissive cells, a fraction of the infected cells may survive and give rise to foci of cells transformed by the virus. In most cases, these transformed cells have integrated into their own genomes viral genes that encode viral transforming proteins, which are capable of modifying cell growth programs (for a review in the case of adenovirus, see Branton et al. 1985). By studying the mechanisms of action of viral transforming proteins, we hope to understand the cellular mechanisms that control proliferation.

The normal, nonviral regulators of cell proliferation in vivo include growth factors that induce signals to cells through binding of the growth factor to specific transmembrane receptors. Among the immediate effects of growth factors is the activation of specific cellular genes, including c-*myc* (Kelley et al. 1983), c-*fos* (Greenberg and Ziff 1984), and genes for other early response proteins (Lau and Nathans 1987). Several of these early response genes encode proteins that bind to DNA and may act as transcription factors that regulate other genes. Ultimately, the growth-factor-induced cascade of gene activity must culminate in processes that directly control the mechanisms of proliferation. A hypothetical example would be the induction of a regulatory protein that is a component of the DNA synthesis machinery.

Functions of the Adenovirus E1A Gene

The E1A proteins are major adenovirus-encoded regulators of viral interactions with the infected host cell. The E1A gene is expressed from the earliest stages of infection and, by virtue of differential splicing, encodes two nuclear proteins, 243 and 289 amino acids long, respectively. Inspection of the amino acid sequences of E1A proteins encoded by several different adenovirus strains has revealed three highly conserved regions, all of which lie within the first exon (Kimelman et al. 1985). Conserved regions 1 and 2 are shared by the 243- and 289-amino-acid proteins, whereas region 3 constitutes the sequences exclusive to the 289-amino-acid proteins. The following properties have been ascribed to the E1A proteins: the ability to activate RNA-polymerase-II-dependent transcription of viral early genes and several cellular genes; stimulation of transcription by RNA polymerase III of the virus-associated (VA) RNA genes; repression of transcription of enhancer-dependent viral and cellular genes; immortalization of primary cells; cooperation with either the adenovirus E1B gene or an activated *ras* gene to transform primary cells fully; induction of the synthesis of growth factors and induction of cellular DNA synthesis (for reviews, see Berk 1986a,b; Moran and Mathews 1987). Because only the 289-amino-acid protein is an efficient activator of RNA polymerase II, region 3 appears to be crucial to the transcription *trans*-activation function. Because both the 243- and 289-amino-acid proteins immortalize primary cells, cooperate with *ras,* and repress enhancers, these functions may be interrelated and may be provided by regions common to the two proteins. It has been suggested that repression of enhancer-dependent genes provides the mechanistic basis for the ability of E1A to induce cellular proliferative functions and to immortalize or transform cells (Velcich and Ziff 1984, 1985; Lillie et al. 1986; Schneider et al. 1987).

Mutants of E1A Deficient in Transcription Regulation

We have examined the properties of *Xho*I linker insertion mutants of the E1A gene expressed from transfected plasmids or reconstructed mutant viruses. The structures of these mutants and the proteins that they encode are indicated in Figure 1. Two mutant plasmids, pSVXL124 and pSVXL214, are of special interest. The first has a mutation that frameshifts the mRNA, thereby eliminating the carboxy-terminal border of con-

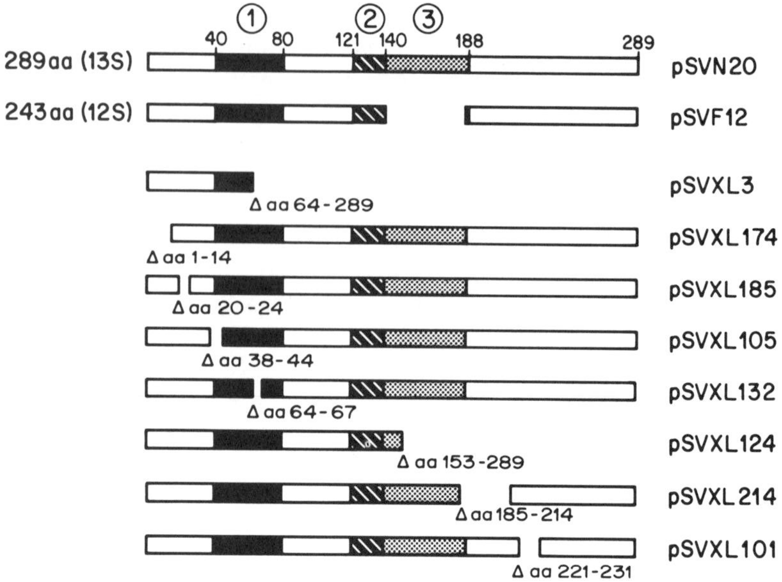

Figure 1 Predicted E1A polypeptide structures encoded by mutant E1A vectors. The 289-amino-acid wild-type protein encoded by the 13S mRNA is represented at the top. The regions that are highly conserved among the adenovirus serotypes are boxed and indicated as domains 1, 2, and 3. The predicted protein structures encoded by the 12S cDNA and the mutant 13S cDNA E1A vectors studied in this paper are illustrated below. The mutations result from the introduction of *Xho*I linkers into the E1A 13S cDNA sequence. Also marked are the wild-type amino acid residues deleted in each mutant.

served region 3 and the following wild-type residues. The other has an in-frame deletion that eliminates the last amino acid residue of conserved region 3, as well as residues within the second exon. Table 1 describes the transcription regulatory properties of several of these mutants assayed as plasmid clones. To measure transcription *trans*-activation, plasmids were transfected into HeLa cells and then superinfected with virus *dl*312, a mutant in which the E1A gene has been deleted. Activation of expression of the gene for the viral DNA-binding protein (DBP) by transfected plasmid-encoded E1A protein was indicative of the *trans*-activation function and was measured by an indirect immunofluorescent assay for DBP (Velcich and Ziff 1985). To measure enhancer repression, the E1A plasmids were cotransfected into HeLa cells, together with plasmid PyE92-1 (de Villier et al. 1982), in which expression of the rab-

Table 1 Transcriptional Properties of Mutant E1A Plasmids

Plasmids	E1A coding sequence	Deletion[a] amino acid	Deletion[a] nucleotide	Transcription *trans*-activation: Cells DBP+ (%)[b]	Transcription repression: Relative level of β-globin mRNA[c]	Deleted domain
PyE92-1 alone	—	—		<0.1	1.00	—
PyE92-1+pSVE1A	genomic	wt	wt	24.9	0.12	wt
PyE92-1+pSVN20	13S	wt	wt	16.8	0.13	wt
PyE92-1+pSVF12	12S	wt	wt	<0.1	0.07	wt
PyE92-1+pSVXL3	13S	64–289	746–753	<0.1	0.81	1,2,3
PyE92-1+pSVXL174	13S	1–14	542–567	6.9	0.24	pre-1
PyE92-1+pSVXL185	13S	20–24	617–632	9.7	0.55	pre-1
PyE92-1+pSVXL105	13S	38–44	671–692	18.6	0.26	pre-1,1
PyE92-1+pSVXL132	13S	64–67	746–761	15.5	0.60	1
PyE92-1+pSVXL124	13S	153–289	1015–1032	1.8	0.90	3, post-3
PyE92-1+pSVXL214	13S	185–214	1110–1319	3.0	0.85	3, post-3
PyE92-1+pSVXL101	13S	221–231	1337–1367	17.9	0.12	post-3

[a]Amino acid residues that are deleted in the mutant protein structures are indicated. As a result of the *Xho*I linker insertion (pCCTCGAGG), linker-encoded amino acid residues replace deleted wild-type (wt) residues. Nucleotide residues (numbering system in Van Ormondt and Galibert [1984]) of the wild-type viral sequences that flank the *Xho*I linker inserted into each mutant are reported.

[b]HeLa cells were transfected with the E1A expression vectors by the $CaPO_4$ precipitate method. *trans*-Activation was measured as the percentage of HeLa cell nuclei immunopositive for DBP. Values are the average of at least two independent experiments. The reported values are the average of at least two independent experiments.

[c]HeLa cells were transfected with 5 μg of PyE92-1 plasmid in the presence or absence of 5 μg of the various E1A vectors. The repression values, obtained by densitometric scanning of the Northern blots, are expressed as the ratio of β-globin mRNA level in the presence of E1A to the level in the absence of E1A DNA, after normalization for the actin mRNA level in each sample. The reported values are the average of at least two independent experiments.

Table 2 *ras* Cooperation Assay in REF52 Cells

Plasmids	Percentage of morphological transformants[a]	Percentage of control (pSVN20)
pT24neo(*ras*) alone	0.6 ± 0.4	2.6
pSVE1A + pT24neo	32.1 ± 12.3	133.8
pSVN20 + pT24neo	24.0 ± 6.8	100.0
pSVF12 + pT24neo	24.7 ± 5.7	102.9
pSVXL3 + pT24neo	0.4 ± 0.4	1.8
pSVXL174 + pT24neo	6.2 ± 3.4	25.8
pSVXL185 + pT24neo	15.0 ± 6.3	62.5
pSVXL105 + pT24neo	17.8 ± 5.5	74.2
pSVXL132 + pT24neo	2.1 ± 1.7	8.8
pSVXL124 + pT24neo	12.5 ± 4.4	52.1
pSVXL214 + pT24neo	23.5 ± 7.6	97.9
pSVXL101 + pT24neo	16.0 ± 4.6	66.7

Different E1A plasmids (1 μg) were transfected with 1 μg of pT24neo.

[a]Cells were fixed and colonies were counted 15–20 days after transfection. The number of morphologically transformed colonies is expressed as average percentage of the total number of G418-resistant colonies from four independent experiments, with the standard error for each sample given.

bit β-globin gene is driven by the polyomavirus enhancer. Enhancer repression lowers the level of β-globin mRNA. We see that the wild-type E1A plasmid, pSVE1A, is an efficient *trans*-activator, conferring expression of DBP on 24.9% of the cells. This plasmid is also an efficient repressor, reducing β-globin mRNA to 12% of the level obtained in the absence of E1A. Note that plasmids pSVXL124 and pSVXL214 lack both the positive and negative transcription regulatory functions.

ras Cooperation Is Not Dependent on Classical RNA Polymerase II Regulatory Functions

We then assayed the capacity of wild-type and mutant E1A genes to cooperate with an activated *ras* gene in transforming REF52 cells, a rat embryo fibroblast cell line that behaves like a primary cell in the *ras* transformation assay (Franza et al. 1986). The results are given in Table 2. The wild-type vector, as well as vectors confined to the expression of either the 289- or 243-amino-acid protein, cooperates efficiently with an activated *ras* plasmid in transforming REF52 cells. Interestingly, pSVXL124 and pSVXL214, which lack transcription regulatory activities, also cooperate efficiently with *ras* in transforming REF52 cells. These results suggest that the *ras* cooperation function of E1A does not depend on a classic RNA polymerase

II regulatory function of E1A. We attempted to identify a protein domain that was responsible for the transcription repression function. Deletions in conserved region 1 (pSVXL 132) and the preceding region (pSVXL185) decrease but do not abolish repression. Repression was lost with pSVXL124 and pSVXL214, plasmids with mutations in conserved region 3 and exon 2. In addition, the work of Lillie et al. (1986) showed that mutations in conserved region 2 eliminated the repressor function. From these results, it appears that no single "domain" of the E1A protein alone accounts for the repressor activity. Perhaps some aspect of the tertiary structure of E1A that is sensitive to mutations at many sites in the protein is required for maintenance of the repressor activity.

Region of E1A Necessary for Induction of Cellular DNA Synthesis

We have also sought to identify the domains of E1A that are required for induction of cellular DNA synthesis and focus formation. We have found that viruses expressing mutant forms of the 243-amino-acid protein, which have substantial deletions overlapping conserved region 1 (e.g., deletion of amino acids 38–67), fail to induce DNA synthesis in primary baby rat kidney (BRK) cells. These mutants also fail to transform BRK cells. Viruses expressing mutants of the 243-amino-acid protein with smaller in-frame deletions (corresponding to plasmids pSVXL185, pSVXL105, and pSVXL132) have a different phenotype. All of these mutants, which have small deletions overlapping conserved region 1 or the preceding sequences, are capable of inducing BRK cellular DNA synthesis, although they do not transform these cells and hence do not generate foci (data not shown). These results suggest that at least two functions are provided by the amino-terminal region of E1A. One is required for induction of cellular DNA synthesis and possibly also focus formation. The other is required for focus formation. We are currently investigating these functions and their relationship to cell-cycle control.

SUMMARY

We have isolated mutants of E1A that are devoid of both the positive and negative RNA polymerase II transcription regulatory functions but can still cooperate efficiently with the *ras* gene in transformation of REF52 cells. The phenotype of these mutants suggests that regulation of RNA polymerase II transcription is not an E1A function that is necessary for *ras* cooperation. We have also shown that E1A conserved region 1 and the preceding region (pre-1) provide at least two distinct functions necessary for viral transformation of BRK cells. One of these appears to be required specifically for the induction of cellular DNA synthesis. This E1A region could interact, hypothetically, with the cellular factors that regulate the onset of cellular DNA replication. We are currently analyzing the mechanisms by which E1A induces cells to traverse the cell cycle and the relationship of these mechanisms to the growth stimulation effects of growth factors.

ACKNOWLEDGMENTS

This work was supported by grants from the National Institutes of Health and the American Cancer Society. A.V. is a special fellow of the Leukemia Society of America. D.S was a fellow of the Sackler Institute of New York University Medical Center.

REFERENCES

Berk, A.J. 1986a. Adenovirus promoters and E1A *trans*-activation. *Annu. Rev. Genet.* **20:** 45.

Berk, A.J. 1986b. Functions of adenovirus E1A. *Cancer Surv.* **5:** 367.

Branton, P.E., S.T. Bayley, and F.L. Graham. 1985. Transformation by human adenoviruses. *Biochim. Biophys. Acta.* **780:** 67.

de Villiers, J., L. Olson, C. Tyndall, and W. Schaffner. 1982. Transcriptional "enhancers" from SV40 and polyoma virus show a cell type preference. *Nucleic Acids Res.* **10:** 7965.

Franza, B.R., Jr., K. Maruyama, J. Garrels, and H.E. Ruley. 1986. In vitro establishment is not a sufficient prerequisite for transformation by activated *ras* oncogenes. *Cell* **44:** 409.

Greenberg, M. and E. Ziff. 1984. Stimulation of 3T3 cells induces transcription of the c-*fos* proto-oncogene. *Nature* **311:** 433.

Kelley, K., B.H. Cochran, C.D. Stiles, and P. Leder. 1983. Cell-specific regulation of the c-*myc* gene lymphocyte mitogens and platelet-derived growth factor. *Cell* **35:** 603.

Kimelman, D., J.S. Miller, D. Porter, and B.E. Roberts. 1985. E1A regions of the human adenoviruses and the highly oncogenic simian adenovirus 7 are closely related. *J. Virol.* **53:** 399.

Lau, L.F. and D. Nathans. 1987. Expression of a set of growth-related immediate early genes in BALB/c 3T3 cells: Coordinate regulation with c-*fos* or c-*myc*. *Proc. Natl. Acad. Sci.* **84:** 1182.

Lillie, J.W., M. Green, and M.R. Green. 1986. An adenovirurs E1a protein region required for transformation and transcriptional repression. *Cell* **46:** 1043.

Moran, E. and M.B. Mathews. 1987. Multiple functional domains in the adenovirus E1A gene. *Cell* **48:** 177.

Schneider, J.F., F. Fisher, C.R. Goding, and N.C. Jones. 1987. Muta-

tional analysis of the adenovirus E1A gene: The role of transcriptional regulation in transformation. *EMBO J.* **6:** 2053.
Van Ormondt, H. and F. Galibert. 1984. Nucleotide sequences of adenovirus DNAs. *Curr. Top. Microbiol. Immunol.* **110:** 73.
Velcich, A. and E. Ziff. 1984. Repression of activators. *Nature* **312:** 594.
———. 1985. Adenovirus E1a proteins repress transcription from the SV40 early promoter. *Cell* **40:** 705.

Cloning of Genes Necessary for G_1 Progression in Animal Cells

C. Basilico, A. Greco, M. Ittmann, and S. Sung Gong

Department of Pathology, New York University School of Medicine
New York, New York 10016

The problem of how gene expression and modulation of function of specific gene products regulate eukaryotic cell proliferation has led several laboratories to attempt genetic analysis of the cell cycle, mainly by isolating and characterizing conditional lethal mutants that arrested growth in specific phases of the cell cycle under nonpermissive conditions. These mutants suggested, particularly in yeast, the existence of specific cell-cycle events whose completion was essential for other steps in the cycle to occur, but, in most cases, the precise gene products and biochemical function involved were not identified. More recently, however, the availability of powerful techniques for gene cloning has allowed the isolation of a number of genes that are essential for and regulate cell-cycle progression. In both *Saccharomyces cerevisiae* and *Schizosaccharomyces pombe* (discussed by Reed et al.; Matsumoto et al.; Draetta and Beach; all this volume), the approach used generally has been to clone genes complementing the defects of *cdc* mutants, because they either code for the wild-type form of the mutated protein or bypass the block (suppressor genes).

In animal cells, two approaches have been mainly followed. One consists of isolating the cDNAs corresponding to genes that are induced in quiescent cells by serum or growth factors (Cochran et al. 1983; Hirschorn et al. 1984; Lau and Nathans 1985; R. Bravo, this volume). This approach, although producing very interesting results, has the disadvantage that the genes identified by this method may not be essential or could play no role in controlling cell-cycle progression. Our laboratory has followed the approach pioneered in yeast and has attempted to isolate genes that complement the defect of temperature-sensitive mutants that are arrested in specific phases of the cell cycle at the nonpermissive temperature (Greco et al.

1987; Ittmann et al. 1987). So far, we have concentrated on G_1 mutants, because G_1 is the "decision" phase of the cell cycle, where cells become committed to DNA synthesis and mitosis.

When comparing the two approaches, one should also take into consideration the questions of whether quiescence (G_0, G_1 arrest, etc.) can be considered a part of the G_1 phase and of whether cells reentering the cycle from G_0 can be considered equivalent to cycling cells in G_1. Although G_0 cells are almost certainly different from cycling cells in G_1, the present state of knowledge does not allow the unequivocal differentiation between G_1-specific and $G_0 \rightarrow G_1$-specific events. In addition, the majority of resting cells in a multicellular organism are arrested in G_1/G_0, and the lack or leakiness of such an arrest is the major difference between the cell-cycle properties of cancer cells and normal cells. Thus, the identification of both G_1-specific and $G_0 \rightarrow G_1$-specific steps is going to be crucial for our understanding of the regulation of cell proliferation in general.

Cloning Strategy

During the last 2 years, our laboratory has isolated and sequenced the human genes complementing the defects of two distinct temperature-sensitive G_1 mutants of the BHK hamster cell line, *ts*11 (Talavera and Basilico 1977) and *ts*BN51 (Nishimoto and Basilico 1978). The strategy used for cloning these genes consisted of transfecting high-molecular-weight human DNA into the temperature-sensitive mutants and selecting for *ts*$^+$ colonies capable of growing at 39.5°C. *ts*$^+$ colonies were tested for the presence of repetitive human DNA sequences of the *Alu*I family by blot hybridization and were thus discriminated from revertants. Successive cycles of transfection with high-molecular-weight DNA from transformants (one in the case of *ts*11, two for *ts*BN51) were performed to dilute out the irrelevant human DNA sequences and obtain a set of *ts*$^+$ transformants that showed common *Alu*I bands upon restriction enzyme digestion and blot hybridization. From the transformant containing the smallest number of *Alu*I fragments, we constructed a genomic DNA library in the EMBL3 bacteriophage λ vector, which was screened for the presence of recombinant phages containing human *Alu*I sequences. Fragments derived from *Alu*I-positive inserts were used as probes to detect the corresponding mRNA in our transformants or human cells. Once an appropriate probe was found, rather than reconstructing a biologically active genomic sequence (this was done later), we used it directly to screen the Okayama-Berg (1982) cDNA expression library. Positive plasmids containing inserts of a size similar to that determined for the corresponding mRNA were tested for the ability to confer to mutant cells the ts^{+} phenotype. Biologically active plasmids were sequenced and further analyzed. A similar approach has been used by Nishimoto et al. (this volume) to isolate the genes complementing two other temperature-sensitive BHK mutants. However, this technique is very time consuming.

When we compare the task facing the investigator trying to isolate animal cell-cycle genes with the situation facing yeast workers, two disadvantages are immediately apparent. First, there are fewer animal cell *cdc* mutants available (because it is much more difficult to isolate them), and they are less well defined (because the events in the cycle are less defined). Second, the cloning procedures used in animal cells so far are much more time consuming. Thus, the danger implicit in this approach to the cell cycle, that of isolating genes that are not "interesting" or whose function is very difficult to elucidate, becomes a much more serious one than in yeast, where (1) one can be more focused in the choice of mutants and (2) one can clone a number of genes more easily and concentrate on the most interesting ones.

To alleviate these problems partially, we tried to determine whether the gene-cloning procedure could be simplified by the direct use of cDNA expression libraries. We transfected the whole Okayama-Berg cDNA expression library (which we knew contained the cDNA capable of complementing the *ts*11 defect) into *ts*11 cells, together with a plasmid marker conferring G418 resistance and selected for ts^{+}, $G418^{R}$ cells. Of four colonies isolated, three contained *ts*11 cDNA sequences in their chromosomal DNA (Table 1), whereas the fourth clone was a revertant, because it did not contain any cDNA plasmids. Because the Okayama-Berg expression vector contains an SV40 origin of replication, we fused the ts^{+} transformants with COS cells (Gluzman 1981), which constitutively produce SV40 large T antigen, and isolated extrachromosomal DNA, whose replication had been induced by SV40 T antigen. Free DNA isolated from the three transformants contained the *ts*11 cDNA sequences. However, these were only a minority of the plasmids rescued. When the rescued DNAs were transfected into bacteria that were then selected for ampicillin resistance, less than 12% of the Amp^{R} colonies contained the *ts*11 cDNA (Table 2) in

Table 1 Transformation of *ts*11 Cells by Transfection with a cDNA Expression Library

Transfected DNA		
pcB7(*neo*)	GM637	Frequency of ts^+, $G418^R$ colonies $\times 10^{-7}$
+	+	5
+	–	<3

*ts*11 cells growing at 33°C were transfected with 10 μg of the GM637 cDNA human expression library, 2 μg of pcB7 plasmid, and *ts*11 carrier DNA up to 30 μg/100-mm plate (containing ~1.5×10^6 cells). Two days after transfection, the cells were replated in several petri dishes and shifted to 39.5°C 1 day later. One week later, the antibiotic G418 (GIBCO, 250 μg/ml) was added. The medium was changed every 3–4 days. Clones were isolated after 2 weeks of selection in G418.

two cases. Therefore, in principle, it should be possible to isolate cDNAs from libraries directly, provided the corresponding mRNAs are not extremely large or are very poorly expressed. From our data, however, it seems unlikely that one could rely on a single cycle of transfection, and probably a second cycle, using high-molecular-weight DNA from the primary cDNA transformants or an amplified pool of rescued plasmids, would be necessary. Thus, this procedure, which should lead to the direct isolation of a biologically active cDNA by fusion rescue of the transformants, should still be faster than the one outlined above and should also allow the use of cDNA libraries from other species; whereas in the case of high-molecular-weight

Table 2 Hybridization of the cDNA Transformants to the *ts*11 Probe

Clone	High m.w. DNA	Pool of rescued plasmid DNA	Individual positive bacterial colonies after transformation with rescued plasmids
1	–	no rescue	—
2	+	+	0/8
3	+	+	0/8
4	+	+	3/8

High-molecular-weight (m.w.) DNAs from ts^+ clones isolated were analyzed by Southern blot hybridization using the *ts*11 cDNA as probe. ts^+, $G418^R$ clones were fused with COS cells by polyethylene glycol. Two days after the fusion, low-m.w. DNA was extracted and analyzed by Southern blot hybridization. An aliquot of the rescued DNA was transfected into HB101 *Escherichia coli*, and bacteria were selected for ampicillin resistance. For each cell line, eight single bacterial colonies were isolated, and their DNA was hybridized to the *ts*11 cDNA probe.

DNA transfection, practically the only DNA that is easily detectable in a rodent DNA background is human DNA.

The *ts*11 and *ts*BN51 Genes

The *ts*11 cDNA sequence predicts a protein of 540 amino acids, which shows no homology with other known proteins so far (Greco et al. 1987). The gene encompasses about 25 kb of human DNA and contains at least 12 exons and 11 introns and an upstream region of about 3 kb that appears to be important for transcription. The gene appears to be expressed in all cells tested so far, in line with its being essential for cell division. The gene maps on the long arm of human chromosome 7.

In an attempt to understand the biochemical function of the *ts*11 protein, we have raised rabbit antibodies against an MS2-polymerase/*ts*11 bacterial fusion protein, or *ts*11 synthetic peptides. For reasons not entirely clear, we have had difficulties in raising high-titer antibodies and only recently have been able to identify conclusively the *ts*11 protein in immunoprecipitates. Figure 1 shows that the *ts*11 protein is apparently 62 kD, suggesting no extensive posttranslational modifications. Further studies should help to elucidate the localization of this protein and its function.

The BN51 gene encodes a protein of 395 amino acids, with a characteristic structure of four hydrophilic domains, consisting of basic and acidic amino acids with partial homology between domains (Ittmann et al. 1987). This structure could indicate that BN51 is a nucleic-acid-binding protein. Immunoprecipitation experiments (Fig. 1) show two main bands of approximately 50 kD and 45 kD that appear to be predominantly cytosolic, although a small amount is also present in the nuclear fraction. We are in the process of determining the exact relationship between these two proteins. The *ts*BN51 gene is not large (~8 kb) but has not been analyzed extensively yet. It maps on human chromosome 8.

It is interesting to note that expression of both *ts*11 and *ts*BN51 is cell-cycle-regulated, or at least proliferation-dependent. Release of mouse 3T3 or BHK cells from a G_1 arrest leads to an induction of *ts*11 mRNA expression, which occurs at about 5–7 hours after induction of proliferation, approximately 6–7 hours before the cells enter S phase (Greco et al. 1987). The increase in mRNA levels is about six- to eightfold and requires protein synthesis. Similar experiments with the BN51 gene indicate a somewhat smaller induction, 3–4 hours

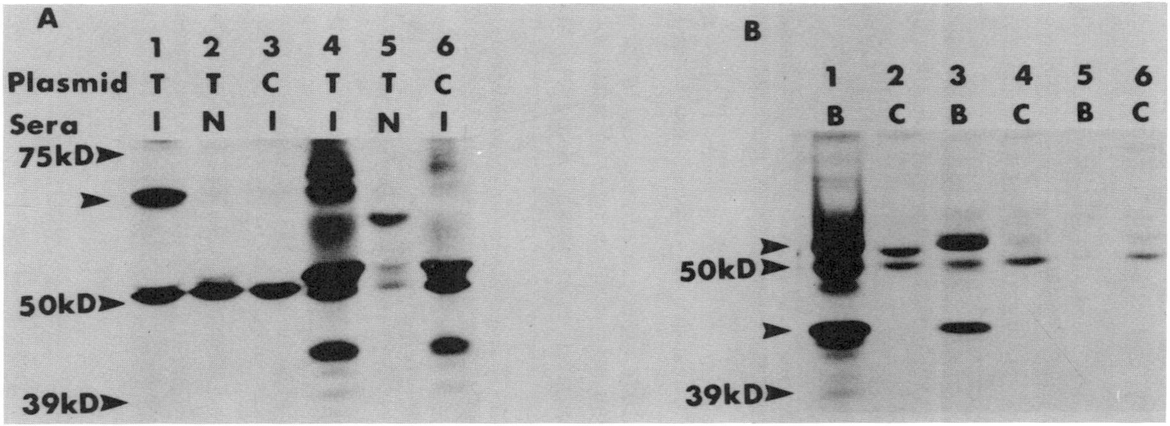

Figure 1 Immunoprecipitation of *ts*11 and *ts*BN51 proteins. COS cells were transfected with plasmids containing either the *ts*11 or *ts*BN51 cDNA inserted in the mammalian expression vector 91023B (Wong et al. 1985) or with control 91023B plasmid, labeled 48 hr later with [^{35}S]methionine and immunoprecipitated. (*A*) SDS-PAGE (10%) of immunoprecipitates from COS cells transfected with the *ts*11 cDNA inserted in the 91023B expression vector (T) or the 91023B vector alone (C). Extracts were immunoprecipitated with either immune (I) or nonimmune (N) sera. (Lanes *1–3*) Sera from rabbit immunized with *ts*11 fusion protein; (lanes *4–6*) sera from rabbit immunized with synthetic peptide corresponding to amino acids 222–234 of the predicted protein sequence of the *ts*11 protein. The position of molecular-weight markers and their sizes (in kilodaltons) are indicated. The relevant band is indicated by an arrow. (*B*) SDS-PAGE (10%) of immunoprecipitates from COS cells transfected with either the BN51 cDNA inserted in the 91023B expression vector (B) or the 91023B vector (C). (Lanes *1* and *2*) Extracts immunoprecipitated with serum from a rabbit immunized with the BN51 fusion protein; (lanes *3* and *4*) extracts immunoprecipitated with antibodies purified from the fusion protein antisera by affinity chromatography, using a column of fusion protein bound to CNBr-Sepharose; (lanes *5* and *6*) extracts immunoprecipitated with nonimmune sera. Molecular-weight markers are as in *A*. The relevant bands are indicated by two arrows.

after serum stimulation. Thus, these genes are not "immediate-early" genes, whose induction by growth factors does not require protein synthesis, and may be expressed in G_1 following some critical event occurring after release from growth arrest. Experiments with other temperature-sensitive mutants are in progress to see whether the expression of these genes is affected by different G_1 blocks.

REFERENCES

Cochran, B.H., A.C. Reffel, and C.D. Stiles. 1983. Molecular cloning of gene sequences regulated by platelet-derived growth factor. *Cell* **33:** 939.

Gluzman, Y. 1981. SV40-transformed simian cells support the replication of early SV40 mutants. *Cell* **23:** 175.

Greco, A., M. Ittmann, and C. Basilico. 1987. Molecular cloning of a gene that is necessary for G_1 progression in mammalian cells. *Proc. Natl. Acad. Sci.* **84:** 1565.

Hirschhorn, R.R., P. Aller, Z. Yvan, C.W. Gibson, and R. Baserga. 1984. Cell-cycle-specific cDNAs from mammalian cells temperature sensitive for growth. *Proc. Natl. Acad. Sci.* **81:** 6004.

Ittman, M., A. Greco, and C. Basilico. 1987. Isolation of the human gene that complements a temperature-sensitive cell cycle mutation in BHK cells. *Mol. Cell. Biol.* **7:** 3386.

Lau, L.F. and D. Nathans. 1985. Identification of a set of genes expressed during the G_0/G_1 transition of cultured mouse cells. *EMBO J.* **4:** 3145.

Nishimoto, T. and C. Basilico. 1978. Analysis of a method for selecting temperature-sensitive mutants of BHK cells. *Somatic Cell Genet.* **4:** 323.

Okayama, H. and P. Berg. 1982. High-efficiency cloning of full-length cDNA. *Mol. Cell. Biol.* **2:** 161.

Talavera, A. and C. Basilico. 1977. Temperature-sensitive mutants of BHK cells affected in cell cycle progression. *J. Cell. Phsiol.* **92:** 425.

Wong, G.G., J.S. Witek, P.A. Temple, K.M. Wilkens, A.C. Leary, D.P. Lexenberg, S.S. Jones, E.L. Brown, R.M. Kay, E.C. Orr, C.S. Shoemaker, D.W. Golde, R.J. Kaufman, R.M. Hewick, E.A. Wang, and S.C. Clark. 1985. Human GM-CSF: Molecular cloning of the complementary DNA and purification of the natural and recombinant proteins. *Science* **228:** 810.

Control of Cell Proliferation in *Saccharomyces cerevisiae* by the Cdc28 Protein Kinase

S.I. Reed, J.A. Hadwiger, M.D. Mendenhall, H.E. Richardson, and C. Wittenberg

Department of Molecular Biology, Research Institute of Scripps Clinic, La Jolla, California 92037

Mutational analysis of the *Saccharomyces cerevisiae* cell-division cycle has allowed the isolation and characterization of mutants defective in division control (Hartwell et al. 1974). One class of such mutants is characterized by a conditional inability to exit from G_1 and initiate new rounds of division (Hartwell et al. 1974; Reed 1980). We have characterized the genes defined by a number of these G_1-arrest mutants and their encoded products in the interest of elucidating the regulatory networks involved in cell-cycle initiation. One such gene, *CDC28*, has been shown both by sequence analysis of the predicted polypeptide and by in vitro assay to encode a protein kinase (Lörincz and Reed 1984; Reed et al. 1985). Presumably, the phosphorylation of target proteins by this kinase is required for cell-cycle initiation, and by inference, regulation of this phosphorylation event may be central to control of proliferation. Furthermore, close structural and functional homologs of the Cdc28 protein kinase have been found in the fission yeast, *Saccharomyces pombe* and, more recently, in human cells, suggesting that the underlying principles of regulation in the system will be general to eukaryotes (Beach et al. 1982; Lee and Nurse 1987).

To confirm that the Cdc28 protein kinase is regulated in response to environmental conditions that normally control cell division, immune complex protein kinase assays were performed on nutrient-limited cells, as well as on cells responding to the mating pheromone α-factor. In each case, protein kinase activity, as defined by the ability to phosphorylate an endogenous 40-kD substrate, was lost (Mendenhall et al. 1987). Furthermore, analysis of synchronous cycling cell populations indicated that active Cdc28 protein kinase was present only during the G_1 interval of the cell cycle (Mendenhall et al. 1987). Thus, activity is restricted to the G_1 interval of actively growing cells, the point where genetic analysis determined that the Cdc28 product must act to initiate cell division. Further analysis of cell-free extracts indicated that Cdc28 activity could be restored to enzyme prepared from non-G_1-cycling cells by adding extracts from *cdc28* mutant cells lacking intrinsic kinase activity but not to enzyme prepared from stationary or pheromone-treated cells. Thus, environmental restraint of cell division causes inactivation of the Cdc28 protein kinase itself, whereas normal exit from G_1 in the course of division results in a loss of activity that can be reversed by the addition of exogenous factors (Mendenhall et al. 1987).

The Cdc28 protein kinase is a phosphoprotein itself (Reed et al. 1985). Therefore, a plausible hypothesis to explain regulation of Cdc28 protein kinase activity is that phosphorylation of the Cdc28 protein by other protein kinases plays a role. We therefore analyzed the phosphorylation state of the Cdc28 polypeptide under conditions associated with active and inactive kinases by two-dimensional gel electrophoresis. No obvious differences in net phosphorylation were detected under any conditions (Hadwiger and Reed 1988). Therefore, it appears that Cdc28 protein kinase activity is not directly regulated at the level of intrinsic phosphorylation. However, we have determined that the active form of the protein kinase is a 160-kD complex of unknown composition and stoichiometry (C. Wittenberg and S.I. Reed, in prep.). This complex is present only during the G_1 interval of actively cycling cells and dissociates when cells are starved or treated with mating pheromones. Thus, it appears that Cdc28 protein kinase activity may be regulated at the level of assembly of a high-molecular-weight active complex (C. Wittenberg and S.I. Reed, in prep.).

In the interest of determining the structure of this protein complex and the regulation of its assembly, we are employing both biochemical and genetic approaches. Purification of the complex is in progress. The genetic strategy of choice is to isolate extragenic suppressors of *CDC28* mutations under the assumption that such suppressors define gene products that must interact with the Cdc28 product either directly or indirectly. Two types of *CDC28* mutations have been used. Suppressors of temperature-sensitive mutations have been obtained by overexpressing random wild-type genes on high-copy plasmids. The rationale for this approach is that temperature-sensitive alleles exhibiting low or labile activity can be rescued by overexpressing other components that promote activity or stabilize an active form of the enzyme. Three sequences have been obtained by this method, designated *PSC1, PSC2,* and *PSC3* (for *p*lasmid *s*uppressor of *cdc28*) (J.A. Hadwiger and S.I. Reed, in prep.). Sequence analysis has revealed that *PSC2* has homology with cyclin A from clam (Swenson et al. 1986), thought to be a mitotic inducer, and that *PSC3* has homology with *SUC1* from *S. pombe* (J.A. Hadwiger and S.I. Reed, in prep.). The latter homology is of particular importance because it has been shown that the *SUC1* product interacts directly with a protein kinase in *S. pombe*, Cdc2, which is highly homologous to the Cdc28 protein kinase (Brizeula et al. 1987). Suppressors of dominant negative mutations are also being employed to identify interactive proteins. Dominant negative alleles of *CDC28* encode inactive protein kinases which, when overexpressed in a wild-type cell, confer a loss-of-function phenotype (Mendenhall et al. 1988). Genetic evidence indicates that the overexpressed dominant negative protein competes for limiting components essential for Cdc28 protein kinase function. Using a strategy similar to the one described above, we have isolated high-copy wild-type sequences that can rescue the dominant negative phenotype (Mendenhall et al. 1988; H.E. Richardson and S.I. Reed, in prep.). Presumably, the rescuing sequences encode proteins important for Cdc28 protein kinase function, such as regulatory subunits or substrates. Three different rescuing sequences have been identified, and their analysis is in progress. It is hoped that genetic and biochemical characterization of proteins encoded by suppressors will help to elucidate functional and structural aspects of the Cdc28 protein kinase complex and its relationship to the process of cell-division control. In particular, we would like to identify the biologically relevant substrates of this protein kinase.

Other work in progress is aimed at characterizing the functions of proteins encoded by other genes defined by G_1-arrest mutants. The ultimate goal of these investigations is the detailed description of the signaling networks involved in division control for a simple eukaryotic organism amenable to genetic manipulation in the hope that it will be possible to generalize to more complex systems.

REFERENCES

Beach, D., B. Durkacz, and P. Nurse. 1982. Functionally homologous cell cycle control genes in fission yeast and budding yeast. *Nature* **300:** 706.

Brizuela, L., G. Draetta, and D. Beach. 1987. p13^{suc1} acts in the fission yeast cell division cycle as a component of the p34^{cdc2} protein kinase. *EMBO J.* **6:** 3507.

Hadwiger, J.A. and S.I. Reed. 1988. Invariant phosphorylation of the *Saccharomyces cerevisiae CDC28* protein kinase. *Mol. Cell. Biol.* (in press).

Hartwell, L.H., J. Culotti, J.R. Pringle, and B.J. Reid. 1974. Genetic control of the cell division cycle in yeast. *Science* **183:** 46.

Lee, M.G. and P. Nurse. 1987. Complementation used to clone a human homologue of the fission yeast cell cycle control gene *cdc2*. *Nature* **327:** 31.

Lörincz, A.T. and S.I. Reed. 1984. Primary structure homology between the product of yeast cell division control gene *CDC28* and vertebrate oncogenes. *Nature* **177:** 753.

Mendenhall, M.D., C.A. Jones, and S.I. Reed. 1987. Dual regulation of the yeast *CDC28*-p40 protein kinase complex: Cell cycle, pheromone and nutrient limitation effects. *Cell* **50:** 927.

Mendenhall, M.D., H.R. Richardson, and S.I. Reed. 1988. Dominant negative protein kinase mutations that confer a G1 arrest phenotype. *Proc. Natl. Acad. Sci.* **85:** (in press).

Reed, S.I. 1980. The selection of *S. cerevisiae* mutants defective in the start event of the cell division cycle. *Genetics* **95:** 561.

Reed, S.I., J.A. Hadwiger, and A.T. Lörincz. 1985. Protein kinase activity associated with the product of the yeast cell division cycle gene *CDC28*. *Proc. Natl. Acad. Sci.* **82:** 4055.

Swenson, K.I., K.M. Farrell, and J.V. Ruderman. 1986. The clam embryo protein cyclin A induces entry into M phase and the resumption of meiosis in *Xenopus* oocytes. *Cell* **47:** 861.

Cell-cycle Control within the G_1 Phase of *Saccharomyces cerevisiae*

K. Matsumoto,[1] I. Miyajima, N. Nakayama,[1] K. Kaibuchi,[1] A. Miyajima,[1] K. Arai,[1] K. Tanaka,[2] A. Toh-e,[2] M. Nakafuku,[3] and Y. Kaziro[3]

[1]Department of Molecular Biology
DNAX Research Institute of Molecular and Cellular Biology
Palo Alto, California 94304

[2]Department of Fermentation Technology
Hiroshima University
Higashihiroshima 724, Japan

[3]Institute of Medical Science
University of Tokyo, Minatoku, Tokyo 108, Japan

The mitotic cycle in the budding yeast *Saccharomyces cerevisiae* is initiated in the G_1 phase of the cell cycle. The G_1 phase of yeast can be classified into early and late stages (Fig. 1) (Pringle and Hartwell 1981). Early G_1 is the point at which yeast cells decide whether to enter the resting stage, G_0, or to commit to a proliferative cycle. Entry into G_0 is determined by nutritional cues. Late G_1 is the point of the cell cycle from which haploid yeast cells enter the mating pathway. Mating pheromones cause cell-cycle arrest of the growing haploid cells at this point in order to initiate the mating program.

Temperature-sensitive *cdc* mutations that block cell division in G_1 have been divided into two classes (Reed 1980). One class of mutants (*cdc25, cdc33,* and *cdc35*) arrest at early G_1 and resemble nutritionally arrested cells, whereas another class (*cdc28, cdc36, cdc37,* and *cdc39*) arrest at late G_1 and resemble mating-pheromone-arrested cells. In this paper, we describe the mechanisms of control in early and late G_1 phases of the yeast cell cycle.

cAMP Pathway at Early G_1 Phase

Genetic and biochemical studies have shown that cAMP plays an important role at early G_1 (Fig. 1) (Matsumoto et al. 1985). The *CDC35* gene is allelic with *CYR1* and codes for the cata-

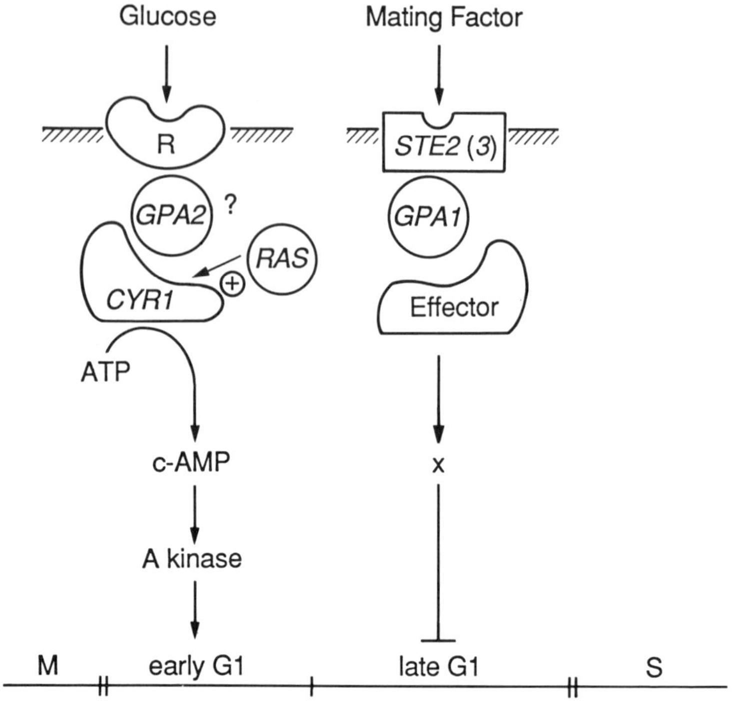

Figure 1 Cell-cycle control within the G_1 phase of *S. cerevisiae*. (Terminal arrowhead) Stimulation; (terminal bar) inhibition; (x) a putative second messenger; (+) activation; (A kinase) cAMP-dependent protein kinase.

lytic subunit of adenylate cyclase, whereas the *CDC25* gene product appears to be involved in positive control of adenylate cyclase activity. The yeast *RAS1* and *RAS2* genes are also involved in the activation of adenylate cyclase (Toda et al. 1985).

Hormonal regulation of adenylate cyclase in mammalian cells is modulated by two G proteins, Gs and Gi (Gilman 1987). Although yeast *RAS* proteins have been considered as functionally equivalent to mammalian Gs protein, we have found that the yeast G protein encoded by *GPA2* is involved in the regulation of cAMP levels (Nakafuku et al. 1988).

GPA2 has been isolated from yeast by cross-hybridization with cDNAs that code for the α-subunit of rat brain Gi and Go. *GPA2* encodes a protein of 449 amino acid residues, with a calculated molecular weight of 50,516. The predicted primary structure of the *GPA2* protein is homologous to mammalian G proteins. To examine the possibility that *GPA2* may be involved in the regulation of cAMP levels, we have studied the effect of *GPA2* on cAMP formation. The *gpa2* disruption mutant is viable and did not affect the cAMP level. On the other hand, a multicopy plasmid carrying *GPA2* elevated levels of cAMP markedly and suppressed the growth defect in a temperature-sensitive *ras2* mutant at high temperature. These results suggest that *GPA2*, in addition to *RAS1* and *RAS2*, regulates the level of cAMP in yeast. However, it is not yet clear at present whether *GPA2* is directly involved in the activation of adenylate cyclase or whether overexpression of *GPA2* indirectly affects the cAMP level.

cAMP exerts its effect through the binding regulatory subunits (*BCY1* gene product) of cAMP-dependent protein kinase, thereby freeing active catalytic subunits. The *bcy1* mutation is defective in the regulatory subunit, thereby allowing the protein kinase to function in the absence of cAMP. Thus, *bcy1* can suppress the growth defect of *cdc25, cdc35/cyr1*, and *ras1 ras2* (Matsumoto et al. 1982; Toda et al. 1985). In addition, early G_1 arrest caused by nutritional limitation does not occur in *bcy1*. These findings demonstrate that cAMP-dependent protein phosphorylation is involved in the early G_1 phase of the cell cycle. To understand how the cell cycle is controlled by cAMP-dependent protein phosphorylation, it is important to identify target proteins of cAMP-dependent protein kinase and their functions. We have found that the expression of the polyubiquitin gene, *UBI4*, is regulated by cAMP, and *UBI4* is required for early G_1 arrest (Tanaka et al. 1988).

When cAMP was depleted, *cyr1* mutants arrested at the early G_1 phase and the *UBI4* mRNA was induced. Because cycloheximide did not inhibit the induction of *UBI4* transcription, de novo protein synthesis was not required for this process. On the other hand, *UBI4* was not induced by the depletion of cAMP in *cyr1 bcy1* double mutants whose cAMP-dependent protein kinase was constitutively active. These results suggest that cAMP-dependent protein kinase regulates the expression of *UBI4* by phosphorylation of positive or negative transcriptional factor(s). Because cAMP depletion caused early G_1 arrest in *cyr1* mutant cells, we wondered whether *UBI4* induction is essential for early G_1 arrest. Phenotypes of *ubi4* disruption mutants support this idea; i.e., *ubi4* disruption mutants are sensitive to nutrient starvation and are defective in sporulation. The *ubi4* disruption mutants cannot arrest at early G_1 phase under nutrient limitation. These results suggest that cAMP depletion controls early G_1 arrest by regulating the synthesis of early G_1-specific gene products, including polyubiquitin.

Mating-pheromone-mediated Signal Transduction at Late G_1 Phase

S. cerevisiae has two different types of haploid cells, α and *a*, which produce specific mating pheromones (called α- and *a*-factor). These mating pheromones interact with specific membrane receptors, *STE2* and *STE3* gene products, and cause cell-cycle arrest of the target cells at late G_1 phase (Sprague et al. 1983). Arrested α and *a* cells then fuse to form diploid cells α/*a*. The *STE2* and *STE3* proteins are predicted to have seven potential transmembrane domains, as for rhodopsin and β-adrenergic receptor, which interact with mammalian G proteins in the signal transduction process (Gilman 1987). Therefore, a G protein may interact with *STE2* and *STE3* proteins to transmit mating-factor signals in a manner analogous to that of mammalian G protein. We have found that the yeast G protein encoded by *GPA1* is involved in mating-pheromone-mediated signal transduction in yeast haploid cells (Fig. 1) (Miyajima et al. 1987).

GPA1 encodes a protein of 472 amino acid residues, with a calculated molecular weight of 54,075, and is expressed only in haploid cells. Disruption of *GPA1* is lethal in haploid cells but not in diploid cells, indicating that *GPA1* is a haploid-specific gene essential for cell growth. The loss of the *GPA1* function leads to cell-cycle arrest at the same stage in the cell cycle as the mating pheromones do. Mutants (*sgp1* and *sgp2*) that suppress the lethality of the *gpa1* disruption mutation show a sterile phenotype that is not cell-type-specific. These mutants produce, but do not respond to, mating pheromones. Loss of the *GPA1* function relieved the sterility caused by the *ste2* disruption mutation. These results suggest that *GPA1* is involved in the mating signal transduction. In this signal transduction, it is important to identify the effector associated with the *GPA1* protein and characterize the nature of the second messenger produced by the effector.

Progression of the early G_1 phase is positively regulated by the nutrient and cAMP cascades, whereas the late G_1 phase is negatively regulated by mating pheromones. Our final objective is to elucidate the sequential biochemical events that ultimately regulate transition from G_1 to S. This knowledge will provide insight into how a set of growth or differentiation factors of mammalian cells transduce growth regulatory signals from the cell surface to the nucleus.

REFERENCES

Gilman, A.G. 1987. G proteins: Transducers of receptor-generated signals. *Annu. Rev. Biochem.* **56:** 615.

Matsumoto, K., I. Uno, and T. Ishiakawa. 1985. Genetic analysis of the role of cAMP in yeast. *Yeast* **1:** 15.

Matsumoto, K., I. Uno, Y. Oshima, and T. Ishikawa. 1982. Isolation and characterization of yeast mutations deficient in adenylate cyclase and cAMP-dependent protein kinase. *Proc. Natl. Acad. Sci.* **79:** 2355.

Miyajima, I., M. Nakafuku, N. Nakayama, C. Brenner, A. Miyajima, K. Kaibuchi, K. Arai, Y. Kaziro, and K. Matsumoto. 1987. *GPA1*, a haploid-specific essential gene, encodes a yeast homolog of mammalian G protein which may be involved in mating factor signal transduction. *Cell* **50:** 1011.

Nakafuku, M., T. Obara, K. Kaibuchi, I. Miyajima, A. Miyajima, H. Itoh, S. Nakamura, K. Arai, K. Matsumoto, and Y. Kaziro. 1988. Isolation of a second G protein homologous gene (*GPA2*) from *Saccharomyces cerevisiae* and studies on its possible function. *Proc. Natl. Acad. Sci.* **85:** 1374.

Pringle, J.R. and L.H. Hartwell. 1982. The *Saccharomyces cerevisiae* cell cycle. In *The molecular biology of the yeast* Saccharomyces cerevisiae: *Life cycle and inheritance* (ed. J.N. Strathern et al.), p. 97. Cold Spring Harbor Laboratory, Cold Spring Harbor, New York.

Reed, S.I. 1980. The isolation of *S. cerevisiae* mutants defective in the start event of cell division. *Genetics* **95:** 561.

Tanaka, K., K. Matsumoto, and A. Toh-e. 1988. Dual regulation of the expression of the polyubiquitin gene by cyclic AMP and heat shock in yeast. *EMBO J.* **7:** 495.

Toda, T., I. Uno, T. Ishikawa, S. Powers, T. Kataoka, D. Broek, S. Cameron, J. Broach, K. Matsumoto, and M. Wigler. 1985. In yeast, *RAS* proteins are controlling elements of adenylate cyclase. *Cell* **40:** 27.

Characterization of *DAF1*, a Mutation Affecting Two Aspects of Start Control in *Saccharomyces cerevisiae*

F. Cross

Fred Hutchinson Cancer Research Center
Seattle, Washington 98104

The Start event in the *Saccharomyces cerevisiae* cell cycle is a point of physiological regulation of cell division. Mating pheromones cause arrest at Start, as does nutritional deprivation; cells that have not reached a critical cell size also do not pass Start (for review, see Pringle and Hartwell 1981). Here, I characterize a mutation (*DAF1*) that affects two aspects of Start regulation simultaneously: pheromone arrest and size control.

RESULTS

DAF1 Cells Are Defective in Division Arrest but Not mRNA Induction in Response to α-factor

DAF1 mutations were isolated as dominant α-factor resistance mutations. The partial (*DAF1* gene-dosage-dependent) resistance phenotype is illustrated in Figure 1.

Figure 2A shows that *STE2* RNA (Burkholder and Hartwell 1985) can be induced approximately normally by α-factor in *DAF1* strains. A dose-response experiment was performed to test the amount of α-factor required to induce *FUS1* mRNA (McCaffrey et al. 1987); similar kinetics were observed for *daf1*$^+$ and a 4x *DAF1* strain (Fig. 2B).

DAF1 Cells Are Small

Cells with five independent alleles of *DAF1* are smaller than wild type. Photomicrographs of *DAF1* and *daf1*$^+$ *a* haploids are shown in Figure 3. Figure 3 also includes cells with deletions of *daf1*$^+$ (created by transplacement; see below) and high-copy *DAF1* integrants. The reduction in cell volume caused by *DAF1* was confirmed by assay with a Coulter channelizer (data not shown). Cell size is monitored at Start, and α-factor arrests cells at Start (Pringle and Hartwell 1981). This pleiotropy suggests that *DAF1* might have a general involvement at Start.

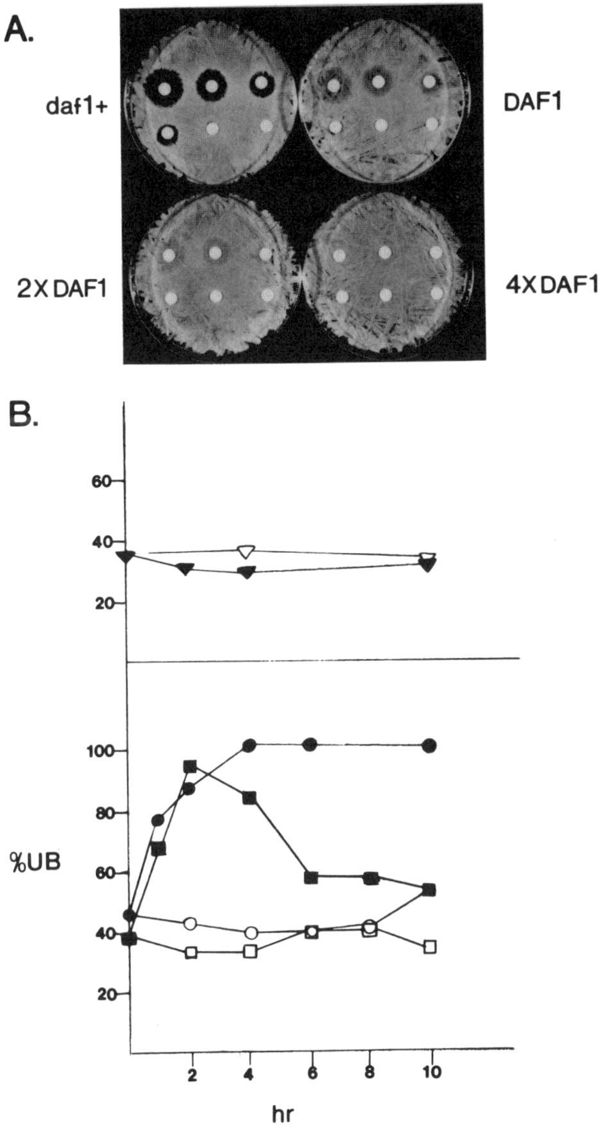

Figure 1 Assays of α-factor resistance of *DAF1* and *daf1*$^+$ strains. (*A*) Lawns of *MATa bar1* strains of the indicated *DAF1* genotype were challenged with threefold serial dilutions of α-factor and incubated overnight. 2x and 4x are multiple-copy integrants of *DAF1*. (*B*) Liquid YEPD cultures in early log phase (30°C) were split, and α-factor was added to half to 10^{-7} M. At intervals, aliquots were removed, fixed, and sonicated, and the percentage of unbudded cells was scored. (Open symbols) No α-factor; (closed symbols) + α-factor. (Circles) *daf1*$^+$; (squares) *DAF1*; (triangles) 4x *DAF1*.

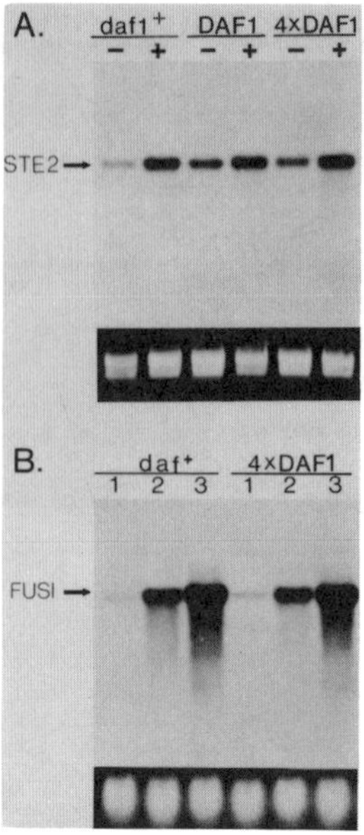

Figure 2 Induction of *STE2* and *FUS1* mRNAs by α-factor. *MATa* strains of the indicated genotype were incubated with different concentrations of α-factor for various times. RNA was isolated and run on formaldehyde-agarose gels and transferred to nylon membranes, which were hybridized with RNA probes to *STE2* (Burkholder and Hartwell 1985) or *FUS1* (McCaffrey et al. 1987). (*A*) Cells were incubated without or with 10^{-7} M α-factor for 3 hr. (*B*) Cells were incubated with 0, 10^{-11}, 10^{-10}, or 10^{-9} M α-factor for 20 min (lanes *1–4*, respectively).

Cloning of *DAF1*

I cloned *DAF1* on the basis of its dominant mutant phenotype by making a YCp50 library from mutant DNA, transforming, and selecting with α-factor. The identity of the cloned gene with *DAF1* was shown by integration into the genome, followed by tetrad analysis of integrants (data not shown) and by the fact that a deletion of the cloned gene transplaced into the genome reverted *DAF1* in *cis* but not in *trans* (see below). The cloned gene was also used to make multiple integrants (referred to as 2x and 4x). Hybridization to *Sfi*I–*Not*I chromosome digests (A. Link, pers. comm.) and published data on cloned regions of chromosome I (Coleman et al. 1986), combined with plasmid mapping data on *DAF1*, resulted in mapping *DAF1* to a 2.2-kb *Eco*RI–*Sal*I fragment, adjacent to *CDC24* (Fig. 4).

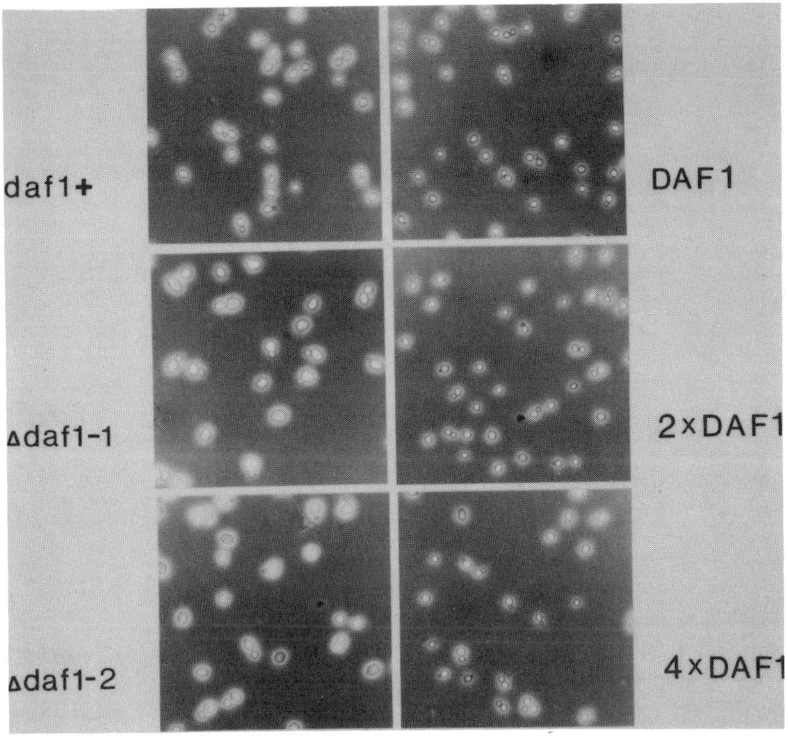

Figure 3 Photomicrographs of *daf1*$^+$, *DAF1*, and *daf1*-deleted strains. Strains of the indicated *DAF1* genotype were grown to log phase in YEPD liquid culture, fixed, sonicated, and photographed under phase-contrast microscopy. *daf1-1* and *daf1-2* are two deletions (see Fig. 4).

Sequence of the *DAF1* Region and Identification of the *DAF1* Mutation

DNA sequence analysis was performed from the *Sal*I site to the *Hpa*I site (Fig. 4), encompassing the *DAF1* region. One large open reading frame was found, which most likely corresponds to the *DAF1* gene. 397 amino acids of open reading frame (ocher terminated) is followed by another 182 amino acids of open reading frame in the same frame (Fig. 4). The *daf1*$^+$ gene was recovered by the gap-repair method (Orr-Weaver et al. 1983; Kunes et al. 1987). Partial sequencing of this gene combined with marker-rescue experiments showed that the first termination codon in *DAF1* was not present in *daf1*$^+$ and was sufficient to convert the *daf1*$^+$-coding sequence to *DAF1*.

The *daf1*$^+$ sequence can be roughly divided into a basic amino-terminal third, an acidic middle third, and a basic

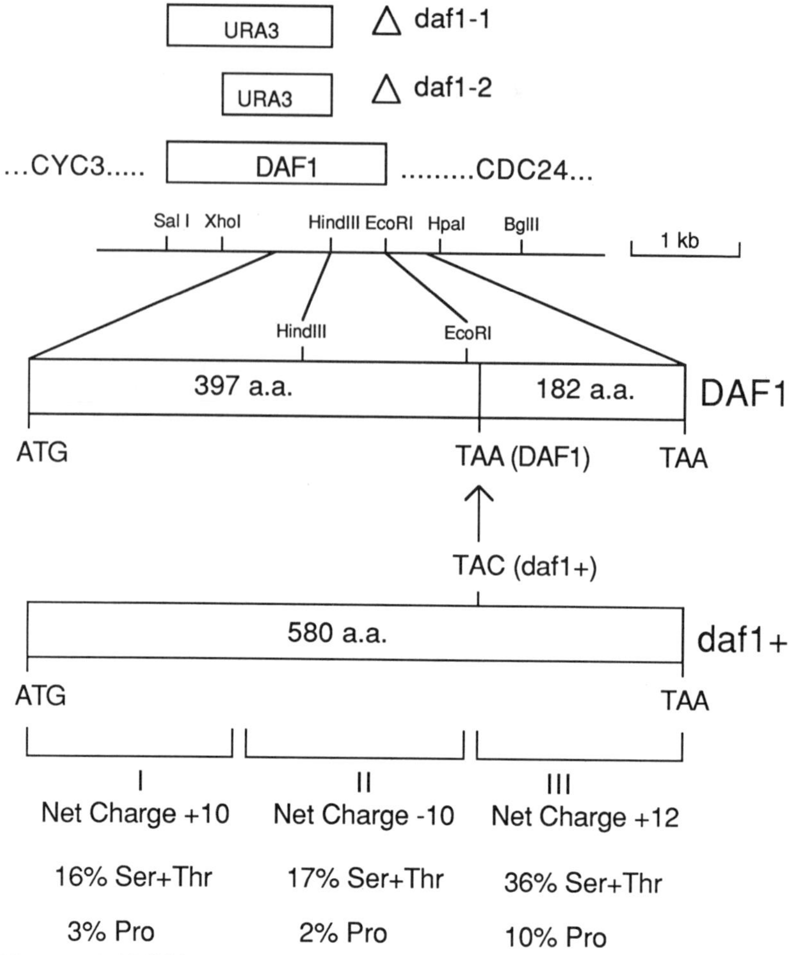

Figure 4 *DAF1* mapping and structure. The restriction map of the *DAF1* region is shown, with the minimum *DAF1* region defined in YCp50 boxed. The locations of *CDC24* and *CYC3* are indicated, as are the locations of the deletions in *daf1-1* and *daf1-2*. The structures of the *DAF1* and *daf1*$^+$ coding regions are diagramed.

carboxy-terminal third, containing a very high proportion of serine and proline residues; the latter region is removed by the *DAF1* truncation.

Deletions of *DAF1* Make Cells Big

Deletions of *DAF1* (see Fig. 4), removing the amino-terminal 237 amino acids and 5′-flanking sequences, resulted in viable cells; these cells were at least 1.5 times the volume of *daf1*$^+$

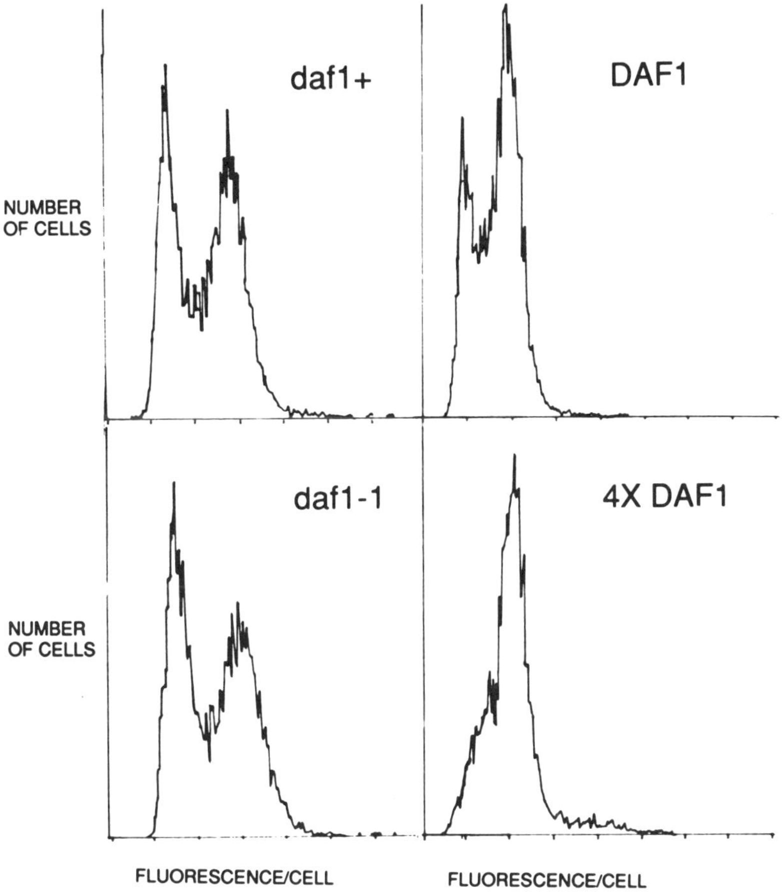

Figure 5 Distribution of DNA content in log-phase cells. Strains of the indicated genotype were harvested to log phase in YEPD at 30°C and processed for flow cytometry.

cells. Transformation with a plasmid containing the *Bgl*II–*Sal*I fragment from the *DAF1* region (Fig. 4) almost completely reverts the large size of cells bearing the deletion. Thus, the increased size of these cells is probably due specifically to the deletion of *daf1*$^+$.

The fraction of unbudded and G_1 cells in log-phase cultures of these strains is slightly increased (Fig. 5; data not shown). Most or all of the increased volume of deletion strains can be accounted for by increased size of the mother cell at budding (Fig. 3). Therefore, these cells may be bigger specifically because they do not pass Start until they reach a cell size larger than the normal critical size (Johnston et al. 1977; Pringle and Hartwell 1981). Assuming that these deletions are null alleles, these results imply that *daf1*⁺ has a role in setting the critical size but that this role is not essential.

DAF1 and Cell-cycle Kinetics

The alterations in cell size caused by various alleles at *DAF1* suggested the possibility that *DAF1* was altering cell-cycle kinetics. Two morphogenetic events of the cell cycle, bud emergence and nuclear division, occurred with fairly similar kinetics in *DAF1* and *daf1*$^+$ cells (data not shown). However, Figure 5 shows that *DAF1* reduced the proportion of G_1 cells, and high-copy *DAF1* strains showed no significant G_1 peak, suggesting that these cells start to synthesize DNA very soon after mitosis.

Interactions with a *cdc28* Mutation

CDC28 is a gene whose product is required for Start (Reed 1980; Pringle and Hartwell 1981). *DAF1* did not suppress the requirement for *CDC28*, but the *cdc28* mutation partially suppressed the α-factor resistance of *DAF1* at 30°C, a temperature at which both *daf1*$^+$ and *DAF1* strains bearing the *cdc28* allele had a near-normal doubling time. *cdc28 daf1-1* double mutants had a semipermissive temperature lower than that of *daf1*$^+$ *cdc28* strains (not shown). These data are consistent with the idea that *DAF1* and *daf1*$^+$ act at the same point in the cell cycle as *CDC28*, at Start.

DISCUSSION

daf1$^+$ appears to be a gene involved in size control at Start. The gain-of-function mutation *DAF1* makes cells small and resistant to division arrest, but not mRNA induction caused by α-factor. This pleiotropy is intriguing because α-factor arrests cells at the same point in the cell cycle (at Start) where cell size is monitored (Pringle and Hartwell 1981).

One model to explain this pleiotropy is that size control and α-factor arrest operate through a pathway that converges on a single Start-essential product. This product could be *daf1*$^+$; if so, *daf1*$^+$ must be a redundant function because it is nonessential. Another possibility is a "bypass model," in which *DAF1* activity initiates the division cycle before cells reach a critical compartment of G_1 in which they are sensitive to α-factor; an additional result of this early Start is small size at budding.

Mutations in two genes have previously been shown to alter

cell size in *S. cerevisiae* (Carter and Sudbery 1980; Sudbery et al. 1980). Like *DAF1*, the *whi1* mutation causes small size in log-phase growth (*whi2* does not). Recently, *whi1* has been found to map very near to *CDC24* and to cause α-factor resistance (B. Futcher, pers. comm.). Thus, it appears likely that *DAF1* and *whi1* are allelic.

Outstanding questions concerning *DAF1* include: How does it work and what does it interact with? Is *DAF1* a hyperactive or deregulated version of $daf1^+$, or are the *DAF1* activities novel ones acquired by mutation? Is $daf1^+$ a redundant but essential function? If so, how can the equivalent gene be identified? Alternatively, if $daf1^+$ performs a dispensable function, does it modulate the activities of essential components?

REFERENCES

Burkholder, A.C. and L.H. Hartwell. 1985. The yeast α-factor receptor: Structural properties deduced from the sequence of the *STE2* gene. *Nucleic Acids Res.* **13:** 8463.

Carter, B.L.A. and P.E. Sudbery. 1980. Small-sized mutants of *Saccharomyces cerevisiae. Genetics* **96:** 561.

Coleman, K.G., H.Y. Steensma, D.B. Kaback, and J.R. Pringle. 1986. Isolation and characterization of the *CDC24* gene and adjacent regions of the chromosome. *Mol. Cell. Biol.* **6:** 4516.

Johnston, G.C., J.R. Pringle, and L.H. Hartwell. 1977. Coordination of growth with cell division in the yeast *Saccharomyces cerevisiae. Exp. Cell Res.* **105:** 79.

Kunes, S., H. Ma, K. Overbye, M.S. Fox, and D. Botstein. 1987. Fine structure recombinational analysis of cloned genes using yeast transformation. *Genetics* **115:** 73.

McCaffrey, G., F.J. Clay, K. Kelsay, and G.F. Sprague, Jr. 1987. Identification and regulation of a gene required for cell fusion during mating of the yeast *Saccharomyces cerevisiae. Mol. Cell. Biol.* **7:** 2680.

Orr-Weaver, T.L., J.W. Szostak, and R.J. Rothstein. 1983. Genetic applications of yeast transformation with linear and gapped plasmids. *Methods Enzymol.* **101:** 228.

Pringle, J.P. and L.H. Hartwell. 1981. The *Saccharomyces cerevisiae* cell cycle. In *The molecular biology of the yeast* Saccharomyces. *Life cycle and inheritance* (ed. J. Strathern et al.), p. 97. Cold Spring Harbor Laboratory, Cold Spring Harbor, New York.

Reed, S.I. 1980. The selection of *S. cerevisiae* mutants defective in the start event of cell division. *Genetics* **95:** 561.

Sudbery, P.E., A.R. Goodey, and B.L.A. Carter. 1980. Genes which control cell proliferation in *Saccharomyces cerevisiae. Nature* **288:** 401.

A Yeast Cell-cycle Gene Product with Cycle-dependent Changes in Localization

K.M. Hennessy and D. Botstein

Department of Biology, Massachusetts Institute of Technology
Cambridge, Massachusetts 02139

Cell-cycle mutants of the budding yeast *Saccharomyces cerevisiae* exhibit clear morphologic evidence of their cell-cycle defect in the light microscope. This has made it possible to isolate a large number of conditional-lethal (temperature- and cold-sensitive) mutants defective specifically in the cell-cycle pathways. However, distinguishing among mutants defective in different stages of the cell cycle by their phenotypes alone has often been problematic. For instance, defects in DNA synthesis and mitosis display essentially the same arrest phenotype: a large bud, usually with a single nucleus in the mother at the bud neck with a short spindle across the nucleus. The phenotype indicates that arrest has occurred prior to mitosis; not surprisingly, then, a large fraction of all cell-cycle mutants show this phenotype.

We have chosen to study a particular group of cell-division-cycle (*cdc*) mutants displaying the large-bud-arrest phenotype because there is genetic evidence that the gene products interact and are likely to form a protein complex in vivo (Moir et al. 1982). Mutations in two separate genes, *cdc45* and *cdc54*, can be suppressed by the mutations in *cdc46*; however, those alleles that suppress the *cdc45* defect do not suppress *cdc54*, and vice versa. Mutations in a fourth gene, *cdc47*, can also suppress the *cdc45* defect but when combined with *cdc54* are lethal. The results shown in Table 1 are typical of those obtained for interacting gene products.

Table 1 Properties of Cell-cycle Mutations

cdc45-1	cold sensitive	
cdc54-1	cold sensitive	
cdc46-1	temperature sensitive	suppressor of *cdc45-1*
cdc46-5	temperature sensitive	suppressor of *cdc54-1*
cdc47-1	temperature sensitive	suppressor of *cdc45-1*

Because it is possible to isolate and manipulate genes in *S. cerevisiae* by complementation, we have been able to clone the *CDC46* gene and have expressed a portion of it in bacteria as a *trpE*-fusion protein. Affinity-purified antibodies against this fusion protein stain the nuclei of nondividing yeast cells by immunofluorescence. However, this nuclear localization changes as cells pass through the cell cycle. Once budding occurs and division has started, there is an apparent exclusion of *CDC46* staining from the nucleus (Fig. 1). This cell-cycle variation does not appear to be due to differences in overall protein levels during the cell cycle. Synchronized cultures do not show any significant variation in total *CDC46* protein levels through the cell cycle. However, separation of arrested cells into nuclear and cytoplasmic fractions shows that cells with nuclear staining (when arrested with α-factor) have *CDC46* protein primarily in the nuclear fractions, whereas cells with no nuclear stain-

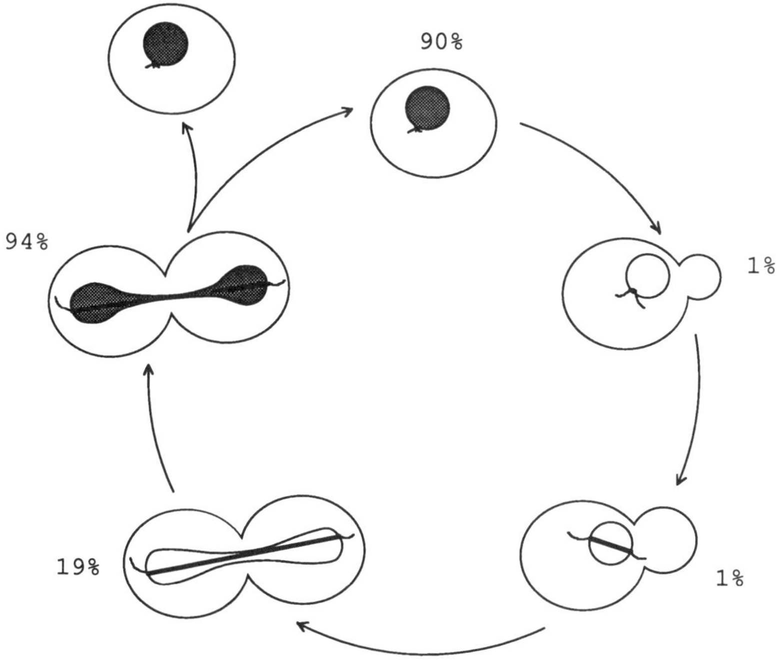

Figure 1 Variation in the *CDC46* protein during the cell cycle. The shading indicates those cell types that typically show *CDC46* nuclear staining. The percentage of cells at each point of the cell cycle (as defined by spindle morphology) that have *CDC46* staining in the nucleus is shown. An asynchronous population of wild-type yeast was double stained with anti-*CDC46* and anti-tubulin antibodies by immunofluorescence.

ing (those arrested with either nocodazole or β-tubulin mutants) have the *CDC46* protein predominantly in the cytoplasmic fractions. The morphological variation therefore seems to be due to differences in antigen localization. Thus, it appears that the *CDC46* protein physically moves into the nucleus as mitosis ends and subsequently moves back into the cytoplasm some time during S phase.

Several assays allow differentiation of these large bud mutations as defects in either DNA synthesis or mitosis. Among these are (1) the frequency of chromosome loss and recombination, (2) determination of DNA content by flow cytometry, and (3) ordering experiments. Both DNA synthesis and mitosis defects lead to high frequency of chromosome loss, whereas a high frequency of recombination is indicative of DNA damage. As expected, tubulin mutants exhibit normal levels of recombination, whereas DNA polymerase mutants have a greatly increased recombination frequency (Hartwell and Smith 1985; Huffaker et al. 1988). Similar analysis of *cdc45* and *cdc46* shows a high frequency of recombination, as well as chromosome loss, which suggests that gaps may be left in the DNA as a result of the *cdc46* defect. Flow cytometry allows the DNA content of individual cells to be determined, using propidium iodide fluorescence to indicate DNA. Arrested tubulin mutants show that DNA synthesis can be completed in the absence of mitosis (Huffaker et al. 1988). Mutants in DNA synthesis, at either the initiation or elongation levels, arrest at the restrictive temperature without DNA synthesis occurring. *cdc46, cdc47, cdc45,* and *cdc54* all arrest prior to DNA synthesis, on the basis of the results of flow cytometry experiments. Thus, these mutants appear to be defective in DNA synthesis.

Because there is reason to suspect that the DNA synthesis pathways have been conserved during evolution, it is likely that protein homologs to *CDC46* protein exist in other eukaryotic cells. In an initial attempt to locate such homologs, we have used affinity-purified antibodies to the yeast *CDC46* protein on both Western blots and for immunofluorescence experiments with a wide range of eukaryotic cells. The *S. cerevisiae CDC46* protein is 90 kD, and the *Schizosaccharomyces pombe* homolog is 80 kD. These same antibodies react with two proteins of 87 and 115 kD in mouse, dog, and human cells. Immunofluorescence staining in both dog kidney and mouse fibroblast cells indicates that the homologous proteins are localized in the nucleus. Thus, it appears that it will be possible to get at the DNA synthesis pathways of other eukaryotic cells through work with *S. cerevisiae.*

The genetic evidence indicates that *cdc45, cdc46, cdc47,* and *cdc54* interact with one another in vivo and are necessary for DNA replication to occur. Whether they are required for initiation or elongation can be determined by ordering experiments. Localization changes in *CDC46* represent a novel cell-cycle control, particularly because there is no nuclear membrane breakdown in this species during the cell cycle.

REFERENCES

Hartwell, L. and D. Smith. 1985. Altered fidelity of mitotic chromosome transmission in cell cycle mutants of *S. cerevisiae. Genetics* **110:** 381.

Huffacker, T., J. Thomas, and D. Botstein. 1988. Diverse effects of β tubulin mutations on microtubule formation and function. *Mol. Cell. Biol.* **160:** (in press).

Moir, D., S. Stewart, D. Osmond, and D. Botstein. 1982. Cold-sensitive cell division mutants of yeast: Isolation, properties, and pseudoreversion studies. *Genetics* **100:** 547.

Regulation of the Expression of the Proliferating Cell Nuclear Antigen and Thymidine Kinase Genes

R. Baserga, B. Calabretta, S. Travali, D. Jaskulski, K.E. Lipson, W.E. Mercer, and J.K. deRiel

Department of Pathology and Fels Research Institute
Temple University Medical School
Philadelphia, Pennsylvania 19140

The proteins of the DNA-synthesizing machinery constitute a group of gene products that are generally expressed coordinately at the G_1/S boundary of the cell cycle (Baserga 1985). We have investigated how growth factors regulate the expression of two of these genes, i.e., that coding for the proliferating cell nuclear antigen (PCNA) and that coding for thymidine kinase (TK). PCNA is a nuclear protein that has been identified recently as the auxiliary protein of DNA polymerase δ (Bravo et al. 1987; Prelich et al. 1987). TK is a member of a group of enzymes involved in the synthesis of cellular DNA. TK enzyme activity (Brent et al. 1965) and TK mRNA steady-state levels (Liu et al. 1985) increase sharply when cells enter S phase. Because of their close association with DNA synthesis, PCNA and TK can be used to study the regulation of expression of an important group of genes that could act as primary elements in the control of cell-cycle progression.

RESULTS

Growth Factor Regulation of PCNA and TK Genes

These studies were carried out in BALB/c-3T3 cells, which are exquisitely sensitive to growth factors. Steady-state mRNA levels were determined in G_0 cells and in cells stimulated by serum or growth factors. Table 1 summarizes the effect of serum and plasma on the mRNA levels of PCNA and TK. The kinetics are slightly different; PCNA mRNA levels are increased by plasma, but the increase is not sustained. TK mRNA is detectable only in cells entering S phase.

Table 1 Steady-state mRNA Levels of PCNA and TK in BALB/c-3T3 Cells

	G_0	8 hr	16 hr	24 hr
PCNA (S)	+	++	+++	++++
(P)		+	–	–
TK (S)	–	–	++	+++
(P)			+	+

G_0 cells were incubated with either 10% fetal calf serum (S) or platelet-poor plasma (P). The amounts are an estimate of Northern blots.

With individual growth factors, the results were as follows: (1) Platelet-derived growth factor (PDGF) alone, or in combination with epidermal growth factor (EGF) and insulin, increased PCNA mRNA levels; (2) PDGF alone was ineffective on TK mRNA levels, which increased only when a combination of growth factors caused the cells to enter S phase; and (3) PCNA, but not TK, was induced by EGF alone.

Coppock and Pardee (1987) have shown previously that low concentrations of cycloheximide (100 ng/ml) completely suppress the induction of TK mRNA by serum. The same conditions also inhibit the increase in PCNA mRNA levels that occur 24 hours after serum stimulation.

Expression of PCNA and TK mRNAs in a G_1-specific Temperature-sensitive Mutant of the Cell Cycle

We used *ts*13 cells (Talavera and Basilico 1977) that arrest in G_1 at the restrictive temperature as the temperature-sensitive mutant. PCNA and TK are regulated similarly in these cells. Both are induced by serum at the permissive temperature; no induction occurs at the restrictive temperature.

Role of the Promoter in the Regulation of *tk* Gene Expression

Four reports have appeared in the literature stating that the cell-cycle regulation of the *tk* gene is determined by the coding sequence and not by the promoter or other regulatory regions (Merrill et al. 1984; Lewis and Markovitch 1986; Hofbauer et al. 1987; Stewart et al. 1987). To test this hypothesis further, we have established cell lines carrying different chimeric constructs of the *tk* gene (Travali et al. 1987). The results are summarized in Table 2. Instead of the term "cell-cycle-regulated" (see Table 2), "growth-factor-regulated" might be more appropriate. This important distinction has been neglect-

Table 2 Regulatory Elements of the *tk* Gene

Promoter	Coding sequence	Cell-cycle-regulated
SV40	TKcDNA	yes
Calcyclin	TKcDNA	no
HSP-70	TKcDNA	no
Human TK	CAT	yes

The appropriate constructs were transfected into cells, and cell lines were established. Gene expression was determined by RNA blots or enzyme activity.

ed so far. Thus, the SV40–*tk* chimeric gene is growth-regulated (Table 2), but its expression is *not* inhibited by cycloheximide, in contrast with the *tk* gene.

Antisense Oligodeoxynucleotides to PCNA Inhibit Cell Proliferation

The *tk* gene codes for a dispensable protein. Cells in culture (and some animals) can live without thymidine kinase. We investigated whether PCNA is necessary for DNA synthesis and cell-cycle progression. For this purpose, we synthesized sense and antisense oligodeoxynucleotides of the first 36 nucleotides of PCNA, which immediately follow the AUG codon. The oligodeoxynucleotides were added to exponentially growing BALB/c-3T3 cells. The antisense oligodeoxynucleotides (but not the sense) effectively inhibit cellular DNA synthesis and cellular proliferation. They also inhibit the expression of the cyclin protein.

DISCUSSION

The three components of the cell cycle that are present in all animal cells are doubling in the size of the cell, DNA replication, and mitosis (Baserga 1985). Evidence is rapidly accumulating that early G_1 events are nonspecific for cell proliferation and that the critical steps for the regulation of cell-cycle progression occur at the G_1/S boundary, when the proteins of the DNA-synthesizing machinery make their appearance. It is reasonable to assume that the regulation of the expression of the genes coding for these proteins holds the key to our understanding of how cell proliferation is regulated. Our investigations are directed at identifying the regulatory elements in two of these genes: PCNA and TK.

REFERENCES

Baserga, R. 1985. *The biology of cell reproduction.* Harvard University Press, Cambridge, Massachusetts.

Bravo, R., R. Frank, P.A. Blundell, and H. MacDonald-Bravo. 1987. Cycling-PCNA is the auxiliary protein of DNA polymerase delta. *Nature* **326:** 515.

Brent, T.P., J.A.V. Butler, and A.R. Carthorn. 1965. Variations in phosphokinase activities during the cell cycle in synchronous populations of HeLa cells. *Nature* **207:** 176.

Coppock, D.L. and A.B. Pardee. 1987. Control of thymidine kinase mRNA during the cell cycle. *Mol. Cell. Biol.* **7:** 2925.

Hofbauer, R., E. Muller, C. Seiser, and E. Wintersberger. 1987. Cell cycle regulated synthesis of stable mouse thymidine kinase mRNA is mediated by a sequence within the cDNA. *Nucleic Acids Res.* **15:** 741.

Lewis, J.A. and T.A. Matkovitch. 1986. Genetic determinants of growth phase-dependent and adenovirus V responsive expression of the Chinese hamster thymidine kinase gene are contained within thymidine kinase mRNA sequences. *Mol. Cell. Biol.* **6:** 2262.

Liu, H.T., C.W. Gibson, R.R. Hirschhorn, S. Rittling, R. Baserga, and W.E. Mercer. 1985. Expression of thymidine kinase and dehydrofolate reductase genes in mammalian S mutants of the cell cycle. *J. Biol. Chem.* **260:** 3269.

Merrill, G.F., S.D. Hauschka, and S.L. McKnight. 1984. TK enzyme expression in differentiation muscle cells is regulated through an internal segment of the cellular TK gene. *Mol. Cell. Biol.* **4:** 1777.

Prelich, G., C.K. Tan, M. Kostura, M.B. Mathews, A.G. So, K.M. Downey, and B. Stillman. 1987. Functional identity of proliferating cell nuclear antigen and a DNA polymerase delta auxiliary protein. *Nature* **326:** 517.

Stewart, C.J., M. Ito, and S.E. Conrad. 1987. Evidence for transcriptional and post-transcriptional control of the cellular thymidine kinase gene. *Mol. Cell. Biol.* **7:** 1156.

Talavera, A. and C. Basilico. 1977. Temperature-sensitive mutants of BHK cells affected in cell cycle progression. *J. Cell. Physiol.* **92:** 425.

Travali, S., K.E. Lipson, D. Jaskulski, E. Lauret, and R. Baserga. 1987. The role of the promoter in the regulation of the thymidine kinase gene. *Mol. Cell. Biol.* **8:** 1551.

Factors Regulating Histone Gene Expression during the Cell Cycle

N. Heintz

Howard Hughes Medical Institute, The Rockefeller University
New York, New York 10021

The transition from the G_1 to the S phase of the eukaryotic cell cycle ultimately results in the activation of DNA synthesis and chromatin replication. It has been established that the transcription of all five classes of histone genes is increased coincidentally with the onset of cellular DNA synthesis. The tight coupling of the rates of histone gene transcription and chromosomal DNA synthesis suggests that the cascade of events leading to entry into S phase may be revealed by the detailed biochemical analysis of either of these processes. Furthermore, it is quite probable that the molecular signals that result in the activation of these mechanisms, and perhaps some of their components, may be identical. Thus, the approach we have taken to understand the transition from G_1 to S phase in the mammalian cell cycle is to define the molecular mechanisms responsible for increased histone gene transcription upon entry into S phase. The key idea, stated simply, is that one can learn the crucial molecular events underlying a given biological transition by backtracking biochemically through a pathway that ultimately results in the activation of specific genes.

The experiments discussed here address two related questions concerning the activation of histone gene expression in S phase: (1) What are the DNA sequences and protein factors involved in the expression of individual mammalian histone genes? (2) How is the coordinate activation of all five histone gene subtypes achieved?

RESULTS AND DISCUSSION

In the past several years, in vitro (Heintz and Roeder 1984; Hanly et al. 1985; Sive et al. 1986; Dailey et al. 1986, 1987; Fletcher et al. 1987) and in vivo (Artishevisky et al. 1985; Artishevisky 1987; Capasso and Heintz 1985; Seiler-Tuyns and

Paterson 1987; Dalton and Wells 1988; LaBella et al. 1988) studies have established that mammalian histone gene expression is regulated by *trans*-acting factors that act through proximal DNA sequences. In this brief review, our most recent studies in this area are described and a working model for cell-cycle regulation of histone gene expression is provided.

Promoter Sequences and Protein Factors Important for Histone H4 Transcription

Our initial studies of the histone H4 promoter first established the complex and distinct properties of mammalian histone gene promoters. To date, at least three upstream sequence elements have been demonstrated to be important for histone H4 gene transcription. The most distal of these elements is located between 82 and 105 nucleotides upstream of the Hu4a cap site and comprises the binding site for the H4TF1 transcription factor. Mutagenesis of this site results in at least a twofold decrease in transcription both in vitro and in vivo. Extensive purification and photo-cross-linking studies indicate that H4TF1 is about 110 kD and that its activity is dependent on the divalent metal cation Zn^{++}. H4TF1 binds to several, but not all, histone H4 promoters. We have not identified binding sites for this protein in other cellular genes, although an exhaustive search for such sites has not been done. Although this protein is clearly functionally active on several histone H4 promoters and may be specific for these promoters, we do not believe it is essential for cell-cycle regulation of histone H4 transcription. Rather, it seems likely that this element and the highly conserved SP1 consensus elements adjacent to it participate in determining the level of transcription and are constitutively active throughout the cell cycle. A possible role for H4TF1 in determining the levels of transcription in different cell types is currently being investigated.

The remaining upstream element in the H4 promoter is the subtype-specific consensus element between 35 and 65 bp upstream of the initiation site. This element is extremely highly conserved in sequence, position, and orientation and is the binding site for transcription factor H4TF2. In this case, purification and photo-cross-linking studies indicate that H4TF2 is a protein of about 60 kD, and its activity is also dependent on divalent metal cations (Zn^{++}, Fe^{++}). As in the case of H4TF1, the purified protein binds avidly to its cognate binding site and it stimulates transcription in a reconstitution assay in a sequence-specific manner. By analogy with the H2B gene (see below), we believe that this protein may play a role in cell-cycle regulation of H4 transcription. However, our initial results with a variety of fusion gene constructs indicate that this may not be the only region of the H4 gene that is required for periodic transcription. Our present efforts are directed toward establishing the exact sequence requirements for regulation of this gene and toward analyzing the behavior of the H4 transcription factors during the cell cycle. To facilitate these analyses, antibodies that react with each of these factors have been prepared and are currently being screened to establish definitively that they interact with the correct DNA-binding proteins.

Transcriptional Regulation of Histone H2B Production during S Phase

Analysis of a variety of 5′ deletion, substitution, and point mutations in the H2B promoter demonstrated that this gene is also controlled by multiple upstream sequence elements and that each of these elements is active in vitro. These elements are arranged in a manner similar to that of the H4 promoter: Immediately upstream of the TATA box is the H2B subtype-specific element (containing the core octanucleotide ATTTGCAT), which is very highly conserved in sequence, position, and orientation; farther upstream in this gene are two separate elements (a set of direct repeats and a CCAAT box), which are less highly conserved. To identify which of the elements is responsible for transcriptional induction upon entry into S phase, each of the mutant promoters was fused to the bacterial chloramphenicol acetyltransferase gene and assayed for its transcriptional phenotype in vivo. The results of these studies can be summarized as follows. Deletion of any sequence element upstream of the H2B subtype-specific element has no effect on the inducibility of the gene upon entry into S phase but can severely reduce the absolute level of transcription. Thus, deletion to position −77 reduces transcription approximately 20-fold, but the residual expression from this promoter is still induced 5-fold upon entry into S phase. In contrast, point mutations that inactivate the H2B subtype-specific element retain wild-type levels of expression prior to S phase but are not induced upon entry into S phase. These results demonstrate that the subtype-specific consensus element is responsible for transcriptional regulation of the H2B gene upon entry into S phase.

Our recent in vitro analysis of these promoter mutants in extracts from synchronized HeLa cells confirmed these in vivo findings and showed that the H2B regulatory element is preferentially active in S-phase extracts. Furthermore, a transcription factor (OTF1), which specifically binds to the H2B octamer sequence and whose functional activity in transcription reconstitution assays is dependent on this sequence, has been purified and shown to be approximately 90 kD. Specific antibodies that react with this 90-kD polypeptide have been prepared and have been used in a series of studies to determine whether the properties of this protein change during the cell cycle. Although these studies are quite preliminary, our initial efforts in this direction have failed to reveal any change in concentration, charge, or DNA-binding characteristics of OTF1 as cells enter S phase.

Promoter Sequences and Protein Factors Involved in Histone H1 Transcription

The organization of the histone H1 promoter is analogous to that of either the H4 or H2B promoters. In this case, however, there are two very highly conserved subtype-specific consensus elements. One of these elements, the H1 "AC box," is located distal in the H1 promoter, and it has been reported that this element is essential for cell-cycle regulation in vivo (Dalton and Wells 1988). We have purified a protein (H1TF1) extensively that specifically interacts with this sequence and believe it to be about 90 kD, although photo-cross-linking or renaturation experiments have not yet definitively shown this to be true. The binding of this protein to the H1 AC box cannot be competed by sequences from any other histone gene promoter. In our studies, deletion of this sequence decreases transcription in vitro or in vivo by approximately twofold.

A second subtype-specific element is present in the H1 promoter immediately upstream of the TATA box, in a position very similar to the analogous elements in the H4 and H2B genes. This 17-bp sequence is very highly conserved in mammalian histone H1 genes and contains within it a CCAAT box. Because it is clear that the CCAAT sequence can be specifically bound by several proteins from HeLa cells, we have carefully characterized a protein that preferentially interacts with the H1 element. This factor (H1TF2) has been shown by photo-cross-linking studies to be 47 kD. Both the sequences required for binding this factor and the DNase footprint of H1TF2 factor on the H1 gene are distinct from previously characterized CCAAT-binding factors (CTF/NF1; Jones et al. 1987). Inactivation of this element in the wild-type promoter results in a severalfold drop in transcription in vivo and in vitro. Our present efforts toward understanding the transcriptional control of H1 genes is to analyze both of the H1 transcription factors further and to acquire specific antibodies to these proteins for studies of their structure and regulation during the cell cycle.

SUMMARY

Our analysis of mammalian histone gene expression during the cell cycle has revealed a surprising complexity in both DNA sequences and protein factors controlling transcription. It is now quite clear that the crucial components for S-phase induction of histone gene expression can be subtype-specific. Thus, neither the DNA sequences nor DNA-binding proteins regulating histone H4, H2B, or H1 expression are shared. However, the general organization of these three promoters is similar and suggests that analogous factors may be performing similar roles in the transcription of these genes. One simple model that incorporates this similarity would be that the subtype-specific domains of these promoters control transcription of these genes during the cell cycle and that the activity of factors binding to these domains is regulated by a single mechanism. We have purified five of these factors and are presently using antibodies raised against them to test this hypothesis.

ACKNOWLEDGMENTS

I acknowledge R.G. Roeder for his involvement in aspects of this work, and I thank the members of both my own and the Roeder laboratories who have participated in these studies for their efforts in understanding histone gene regulation. I am grateful to the Pew Scholars program for their support during the course of these studies.

REFERENCES

Artishevisky, A. 1987. Cell-cycle regulatory sequences in a hamster histone promoter and their interactions with cellular factors. *Science* **328:** 823.

Artishevisky, A., A. Grafsky, and A. Lee. 1985. Isolation of a sequence capable of conferring cell cycle regulation to a heterologous gene. *Science* **230:** 1061.

Capasso, O. and N. Heintz. 1985. Regulated expression of mammalian histone H4 genes in vivo requires a *trans*-acting transcription factor. *Proc. Natl. Acad. Sci.* **82:** 5622.

Dailey, L., S.B. Roberts, and N. Heintz. 1987. RNA polymerase II transcription factors H4TF-1 and H4TF-2 require metal to bind specific DNA sequences. *Mol. Cell. Biol.* **7:** 4582.

Dailey, L., S.H. Hanly, R.B. Roeder, and N. Heintz. 1986. Distinct transcription factors bind to two regions of the histone H4 promoter. *Proc. Natl. Acad. Sci.* **83:** 7241.

Dalton, S. and J.R.E. Wells. 1988. A gene specific promoter element is required for optimal expression of the histone H1 gene in S-phase. *EMBO J.* **7:** 49.

Fletcher, C., N. Heintz, and R.G. Roeder. 1987. Purification and characterization of OTF-1, a transcription factor regulating cell cycle expression of a human histone H2B gene. *Cell* **51:** 773.

Hanly, S.M., G.C. Bleeker, and N. Heintz. 1985. Identification of promoter elements necessary for transcriptional regulation of a human histone H4 gene in vitro. *Mol. Cell. Biol.* **5:** 380.

Heintz, N. and R.G. Roeder. 1984. Transcription of histone genes in extracts from synchronized HeLa cells. *Proc. Natl. Acad. Sci.* **81:** 1713.

Jones, K.A., J.T. Kadonaga, P.J. Rosenfeld, T.J. Kelly, and R. Tjian. 1987. A cellular DNA-binding protein that activates eukaryotic transcription and DNA replication. *Cell* **48:** 79.

Seiler-Tuyns, A. and B.M. Paterson. 1987. Cell cycle regulation of a mouse histone H4 gene requires the H4 promoter. *Mol. Cell. Biol.* **7:** 1048.

Sive, H.L., N. Heintz, and R.G. Roeder. 1986. Multiple sequence elements are required for maximal in vitro transcription of a human histone H2b gene. *Mol. Cell. Biol.* **6:** 3329.

Cellular Proteins Required for Multiple Stages of SV40 DNA Replication In Vitro

B. Stillman, M.P. Fairman, G. Prelich, S. Smith, and T. Tsurimoto

Cold Spring Harbor Laboratory, Cold Spring Harbor, New York 11724

The processes of cell growth, DNA replication, and cell division are highly coordinated and controlled events that are subject to a variety of regulatory influences. During each cell cycle in growing cells, the DNA and chromatin structure must be accurately duplicated once, and only once. Consequently, there are a number of critical regulatory steps ensuring that this occurs without a hitch. The commitment to begin a round of DNA replication is the first critical event in this process; however, once the S phase is under way, other mechanisms must ensure that replication from the multiple origins on cellular chromosomes occurs coordinately but that reinitiation is prevented. Additionally, there must be mechanisms that coordinate elongation from origins that are activated at different times in S phase and a mechanism to ensure that all of the DNA is duplicated before cell division can occur. To understand these events, it is necessary to determine the mechanism of DNA replication in eukaryotic cells and to identify the cellular proteins that may be targets for these regulatory events.

SV40 DNA Replication In Vitro

We have utilized a cell-free replication system that supports the replication of plasmids containing the SV40 origin of DNA replication (Li and Kelly 1984; Stillman and Gluzman 1985). Replication in vitro is dependent on the addition of purified SV40 T antigen (TAg) to extracts prepared from human 293 cells, an adenovirus-transformed human cell line that supports the replication of SV40 DNA in vivo. Replication in vitro occurs bidirectionally from the origin yielding monomer, relaxed daughter molecules after one round of DNA replication. If a nuclear extract is included in the replication reaction, the replicated DNA is assembled into a chromatin structure resembling the structure of cellular chromatin (Stillman 1986). The virus-encoded initiator protein, SV40 TAg, binds to two binding sites within the origin region (I and II); however, only binding site II is essential for initiation of replication. In addition to its specific binding to sites within the origin region, TAg has a DNA helicase/ATPase activity that supports the unwinding of duplex DNA (Stahl et al. 1986; Dean et al. 1987; Goetz et al. 1988). Indeed, TAg, in the presence of a single-stranded DNA-binding protein from *Escherichia coli* (SSB) or HeLa cells, will support the bidirectional unwinding of DNA from the SV40 origin region in the absence of DNA replication (Dean et al. 1987; Dodson et al. 1987; Wold et al. 1987; Goetz et al. 1988).

To identify and characterize the cellular replication factors required for SV40 DNA replication in vitro, we have fractionated the replication extract into multiple chromatographic fractions and reconstituted the cell-free system with purified or partially purified proteins. Following this strategy, a number of cellular factors have been identified and characterized and are described below. As can be seen, we have identified proteins involved in both the initiation and elongation of replication and have already identified a protein that is synthesized in a cell-cycle-dependent manner.

Proliferating Cell Nuclear Antigen/Cyclin

The first cellular protein identified in this laboratory that is required for SV40 DNA replication in vitro was purified from an isolated fraction that bound to DEAE-cellulose and was separated from other essential replication factors. The purified replication factor was a 36,000-dalton (36K) protein which, by a number of criteria, including protein sequence analysis, was identical to a previously characterized protein called the proliferating cell nuclear antigen (PCNA) (Prelich et al. 1987a). PCNA was first identified in 1978 as a human autoantigen, using sera from a small percentage of patients with systemic lupus erythematosis (SLE), and was found in proliferating, but not nondividing cells (for review, see Mathews 1988). Subsequently, PCNA was shown to be identical to a growth- and cell-cycle-regulated protein called cyclin that was preferentially synthesized just prior to the S phase of each cell cycle. Using the human autoantisera, immunofluorescent staining revealed that PCNA localized to discrete nuclear regions in S phase, and the immunofluorescent pattern closely resembled the sites of ongoing cellular DNA synthesis (see Mathews 1988). Thus, our finding that PCNA was required for SV40 DNA replication in vitro strongly implied that it had a similar role in normal cellular DNA replication.

The role played by PCNA in the DNA replication process was suggested by the demonstration that PCNA was identical to another cellular protein called the polymerase δ auxiliary protein (Bravo et al. 1987; Prelich et al. 1987b). This polypeptide stimulates the activity of polymerase δ, but not polymerase α, by increasing the processivity of the polymerase on primed single-stranded DNA templates (Tan et al. 1986; Prelich 1987b), suggesting that PCNA should function during the elongation stage of DNA replication. This was demonstrated by studying the structure of replication products synthesized in reconstituted replication reactions with or without PCNA (Prelich and Stillman 1988). This study not only confirmed the role of PCNA in the elongation process, but also demonstrated that leading-strand synthesis at the replication fork was absolutely PCNA-dependent. Lagging-strand synthesis still occurred in the absence of PCNA, although the short "Okazaki fragment"-sized products were not ligated together.

Overall, these findings suggest that initiation of DNA replication and the synthesis of the first nascent strands at the replication origin occur in a PCNA-independent manner, but there is a switch to a PCNA-dependent stage of replication where PCNA is required for leading-strand DNA synthesis and for coordinating synthesis of both strands at the fork. Because PCNA is a processivity factor for polymerase δ, these results also suggest that DNA polymerase δ is required for leading-strand replication. We also suggest that the DNA polymerase α/DNA primase complex is the lagging-strand polymerase, as well as the polymerase that synthesizes the first nascent strands at the replication origin.

A Mammalian Single-stranded DNA-binding Protein

Another cellular protein, purified on the basis that it is absolutely required for SV40 DNA replication in vitro, is a multisubunit factor containing polypeptides of 70,000, 34,000, and 11,000 daltons in approximate 1:1:1 stoichiometry (Fairman and Stillman 1988). This replication factor, called RF-A, binds to single-stranded DNA with 30- to 50-fold higher affinity than to double-stranded DNA and may function as a mammalian cell single-stranded DNA-binding protein (SSB), analogous to the *E. coli* SSB. The three polypeptides are tightly associated in a complex that is resistant to denaturation with urea and thus probably functions in the cell as such. We are currently determining the role that this protein plays in initiation of DNA replication and whether it has any regulatory role throughout the cell cycle.

Presynthesis Complex Protein

A third cellular protein, tentatively labeled SSI, is essential for the initiation of SV40 DNA replication in vitro and has been isolated, although not yet purified (Fairman and Stillman 1988). However, this replication factor is of considerable interest because it is the only cellular factor that is required to form a presynthesis complex at the replication origin with SV40 TAg. This factor, together with TAg and the plasmid DNA, when incubated for 15 minutes at 37°C, will allow DNA replication to proceed immediately upon addition of the complete complement of replication factors. In contrast, if this presynthesis factor is not preincubated with TAg and the template DNA, but is added with the other cellular components, replication is delayed for 15 minutes at 37°C. This presynthesis step, which defines an additional stage in SV40 DNA replication, requires ATP and an incubation temperature of 37°C, because 30°C will not suffice. Because it is possible that this interesting factor is required for a similar stage prior to initiation of chromosomal DNA replication, we are well on the way toward a full characterization of this protein.

Other Cellular Factors

Further fractionation of the replication extracts has enabled us to confirm the role of topoisomerases I and II during the elongation stage of SV40 DNA replication, but surprisingly, the topoisomerases appear not to be required for initiation of DNA replication. Another cellular protein, which has been purified extensively, is required for the elongation of DNA replication and, as for PCNA, may also be required preferentially for leading-strand, but not lagging-strand, replication.

This analysis, although not complete, has already allowed us to define multiple stages of SV40 DNA replication in vitro. Some of these steps may be analogous to the multiple stages expected for cellular chromosomal DNA replication. It is likely that some of these stages of replication and the components involved will be targets for regulatory pathways that control events such as the temporal control of DNA replication throughout the cell cycle, the regulation of initiation and prevention of reinitiation of replication at the origin, and perhaps also elongation from initiated replication complexes at different times in S phase. Finally, we must ultimately understand how these replication proteins are controlled following stimulation of quiescent, nongrowing cells to proliferate.

REFERENCES

Bravo, R., R. Frank, P.A. Blundell, and H. MacDonald-Bravo. 1987. Cyclin/PCNA is the auxiliary protein of DNA polymerase-δ. *Nature* **326:** 515.

Dean, F.B., P. Bullock, Y. Murakami, C.R. Wobbe, L. Weissbach, and J. Hurwitz. 1987. Simian virus 40 (SV40) DNA replication: SV40 large T antigen unwinds DNA containing the SV40 origin of replication. *Proc. Natl. Acad. Sci.* **84:** 16.

Dodson, M., F.B. Dean, P. Bullock, H. Echols, and J. Hurwitz. 1987. Unwinding of duplex DNA from the SV40 origin of replication by T antigen. *Science* **238:** 964.

Fairman, M.P. and B. Stillman. 1988. Cellular factors required for multiple stages of SV40 replication *in vitro*. *EMBO J.* **7:** 1211.

Goetz, G.S., F.G. Dean, J. Hurwitz, and S.W. Matson. 1988. The unwinding of duplex regions in DNA by the simian virus 40 large tumor antigen-associated helicase activity. *J. Biol. Chem.* **263:** 383.

Li, J.J. and T.J. Kelly. 1984. Simian virus 40 DNA replication in vitro. *Proc. Natl. Acad. Sci.* **81:** 6973.

Mathews, M.B. 1988. The proliferating cell nuclear antigen, PCNA, a cell cycle regulated DNA replication factor. In *Growth control during cell aging* (ed. H.R. Warner and E. Wang). CRC Press, Boca Raton, Florida. (In press.)

Prelich, G. and B. Stillman. 1988. Coordinated leading and lagging strand synthesis during SV40 DNA replication requires PCNA. *Cell* **53:** 117.

Prelich, G., M Kostura, D.R. Marshak, M.B. Mathews, and B. Stillman. 1987a. The cell-cycle regulated proliferating cell nuclear antigen is required for SV40 DNA replication *in vitro*. *Nature* **326:** 471.

Prelich, G., C.-K. Tan, M. Kostura, M.B. Mathews, S.G. So, K.M. Downey, and B. Stillman. 1987b. Functional identity of proliferating cell nuclear antigen and a DNA polymerase-δ auxiliary protein. *Nature* **326:** 517.

Stahl, H., P. Dröge, and R. Knippers. 1986. DNA helicase activity of SV40 large tumor antigen. *EMBO J.* **5:** 1939.

Stillman, B. 1986. Chromatin assembly during SV40 DNA replication *in vitro*. *Cell* **45:** 555.

Stillman, B.W. and Y. Gluzman. 1985. Replication and supercoiling of SV40 DNA in cell free extracts from human cells. *Mol. Cell. Biol.* **5:** 2051.

Tan, C.K., C. Castillo, A.G. So, and K.M. Downey. 1986. An auxiliary protein for DNA polymerase-δ from fetal calf thymus. *J. Biol. Chem.* **261:** 12310.

Wold, M.S., J.J. Li, and T.J. Kelly. 1987. Initiation of simian virus 40 DNA replication *in vitro*: Large-tumor-antigen and origin-dependent unwinding of the template. *Proc. Natl. Acad. Sci.* **84:** 3643.

Positive Control of DNA Replication in the Eukaryotic Cell Cycle: Induction of a Replication Origin Unwinding Activity at the G_1/S Boundary

J. Roberts and G. D'Urso

Fred Hutchinson Cancer Center, Seattle, Washington 98104

To study the positive factors that regulate the onset of DNA replication, we have studied the replication of SV40 DNA in vitro using T-antigen-supplemented extracts prepared from cells at defined positions in the cell cycle. We have found that the abundance of many factors necessary to replicate DNA does not change during the cell cycle. However, the G_1 cell is incapable of using these factors to replicate DNA efficiently. At a discrete point in the cell cycle, just prior to entry into S, an activity appears that is required for site-specific initiation at the SV40 origin. This activity is necessary to unwind DNA at the replication origin.

To begin to understand the events that regulate progression through the cell cycle, we have concentrated on one aspect of this process—the cell's decision to start DNA replication. Genetic and biochemical evidence suggests that at some point in G_1, the cell assesses an array of specific intra- and extracellular cues and decides whether to commit itself to DNA replication or to withdraw from the cell cycle and enter a quiescent state called G_0. Using a cell-free replication assay, we have investigated the molecular basis of the decision to begin DNA replication.

It is important to distinguish the problem we are addressing, the G_1 to S transition, from the mechanisms leading to the cell's return to the cell cycle from the resting G_0 state, the G_0 to S transition. It is not necessarily true that events that follow release from G_0 are normal cell-cycle events that have counterparts in the G_1 to S pathway. The G_0 cell is in a uniquely quiescent state in which most cellular activities necessary for growth and division are suppressed (see Baserga 1984); many gene products that are induced following stimulation of growth-arrested cells are present at constant levels during the normal cell cycle (see Thompson et al. 1985, 1986). Cell-cycle-dependent changes in the abundance of factors necessary for genome duplication have been documented in only a very limited number of cases. Dihydrofolate reductase and histone gene transcription increase at the G_1/S boundary, and thymidine kinase activity is similarly cell-cycle-dependent. In yeast, many factors involved in DNA replication are induced at Start; however, none of the factors analyzed thus far are limiting for the onset of DNA synthesis. The induction of these factors is not necessary for entry into S (B. Beyers, pers. comm.). It has not been determined whether any of the factors in mammalian cells that are induced during DNA replication are rate limiting for the onset of S in the normal cell cycle.

In Vitro Replication of DNA in G_1- and S-phase Cell Extracts

A cell-free system that will faithfully replicate DNA molecules containing the SV40 replication origin has been developed recently (Li and Kelly 1984; Stillman and Gluzman 1985). This system utilizes exogenous SV40 T antigen as the initiator protein and a cytoplasmic extract from primate cells that supplies all the other factors necessary for DNA replication. Because this system relies upon cellular factors for all events in DNA replication other than those performed by T antigen, we anticipated that it would prove useful in assessing cell-cycle-associated changes in their activity.

Our initial experiments were designed to determine whether this cell-free replication system would be a successful indicator of cyclic changes in the abundance of limiting replication factors in human cells. Therefore, we determined the ability of extracts prepared from cells at different positions in the cell cycle to support SV40 replication in vitro. To avoid drug-induced artifactual changes in the abundance of replication factors, we chose to fractionate cells within the cell cycle by centrifugal elutriation.

The G_1- and S-phase extracts from three different cell types were each used to replicate in vitro a plasmid containing the complete SV40 replication origin. We quantitated the level of replication supported by the cell extract by measuring incorporation of radiolabeled dCTP into DNA. The results obtained with G_1- and S-phase extracts from the three cell types were strikingly similar. The specific activity (picomoles of dCTP incorporated per microgram of protein per hour) of S-phase extracts was eight- to tenfold higher than those of G_1 extracts (Fig. 1). Because a G_1 cell contains about half as much total protein as an S-phase cell, the difference in replication activity on a per cell basis was about 20-fold.

In an extension of these experiments, we asked if the ability to replicate DNA increased gradually through G_1 or whether the increased activity suddenly appeared at a discrete point in the cell cycle. Exponentially growing Manca cells were treated for no more than 10 hours with Nocodazole, a reversible inhibitor of microtubule polymerization, to synchronize them in mitosis just prior to metaphase. Ninety minutes following release from the mitotic block, approximately 50–75% of the population had divided and were G_1 cells. At this point, the G_1 cells were separated from residual mitotic cells by centrifugal elutriation. FACS analysis confirmed that this procedure yielded a pure population of G_1 cells. These G_1 cells were returned to culture, and extracts were prepared hourly as they progressed synchronously through G_1 and into S phase. The movement of these cells through the cell cycle was monitored both by measuring incorporation of ^{3}H-labeled dT into DNA and by FACS analysis of DNA content. Our results showed that the ability of extracts prepared from these cells to replicate SV40 DNA in vitro suddenly increased 6 hours after mitosis, just as the cells entered S phase. Therefore, we have identified a discrete point in the cell cycle when the cell extracts become competent to replicate DNA.

Many Factors Necessary to Replicate DNA Are Constitutively Present Throughout the Cell Cycle

We considered three alternative hypotheses to explain the difference in the replicative activity of G_1- and S-phase extracts: (1) G_1 cells contain an inhibitor of DNA replication. (2) G_1 cells are deficient in most, if not all, replication factors. Most factors necessary to replicate DNA are synthesized very late in G_1, just as the cell decides to enter S phase. (3) Most factors necessary

Figure 1 DNA replication in extracts prepared from cells at defined cell-cyle positions. Extracts were prepared from G_1- and S-phase Manca (*A*), 293 (*B*), or K562 (*C*) cells. In addition, G_2 extracts were prepared from K562 cells. Reactions were carried out using 1 μg of T antigen, 100 ng pSV-ori, and the indicated amount of cytoplasmic extract for 1 hr. Incorporation of dCMP into DNA was used to measure the extent of the reactions.

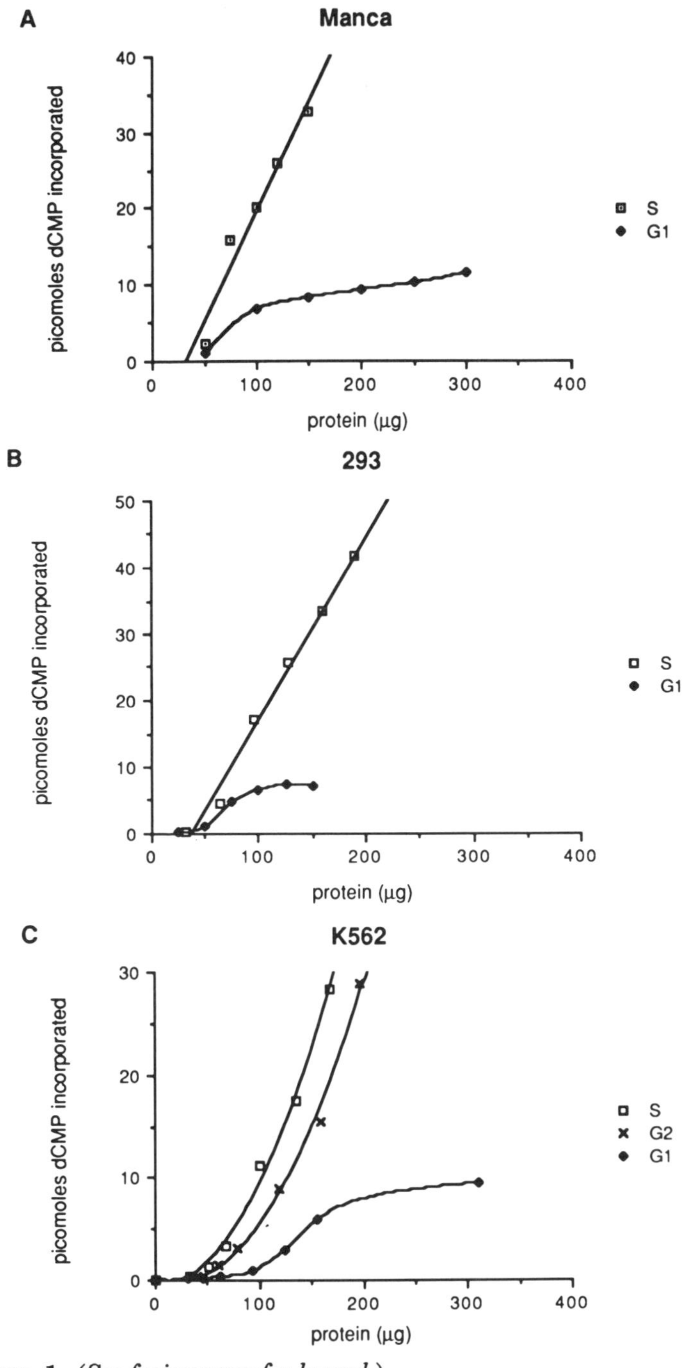

Figure 1 (*See facing page for legend.*)

to replicate DNA are present throughout G_1 and S. The decision to enter S phase involves the synthesis of just one or a few limiting factors that regulate key steps in DNA replication.

Mixing of G_1- and S-phase extracts demonstrated that G_1 cell extracts did not contain an inhibitor of DNA replication. We established assay conditions where the extent of replication was linearly dependent on the amounts of added cytoplasmic extract. As above (Fig. 1), the specific activity of G_1 extracts was about tenfold lower than that of S-phase extract. We then added increasing amounts of G_1 extract to a fixed amount of S-phase extract. The specific activity of the mixed S + G_1 reaction was equivalent to the control S reaction. In contrast, the addition of heat-inactivated G_1 extract or bovine serum albumin (BSA) to the same amount of S-phase extract did not stimulate replication (data not shown). Therefore, the G_1 extract was not neutral but must have supplied replication factors in the G_1 + S mixture, and the S-phase cell extract contained an excess of the factor(s) that was limiting DNA replication in G_1 extracts. Furthermore, the preexisting G_1 replication machinery was responsive to this rate-limiting factor.

In further experiments, we showed that the abundance of three enzymes necessary for DNA replication in this system, topoisomerase I and α plus δ DNA polymerases, was identical in G_1- and S-phase extracts. Also, we found that the addition of purified proliferating cell nuclear antigen or topoisomerase I did not stimulate replication in G_1 extracts, suggesting that neither of these activities was limiting. Finally, visualization of reaction products on agarose gels demonstrates that neither topoisomerase II nor DNA ligase was limiting in G_1 extracts. In sum, as suggested by the mixing experiment, many replication factors were equally abundant in G_1- and S-phase extracts. Our data were most consistent with the hypothesis that only some replication factors were limiting in G_1 extracts and present in excess in the S-phase cell.

A Replication Origin Unwinding Activity Is Induced as Cells Enter S Phase

Kinetic experiments indicated that once initiated, replication proceeded at equal rates on individual DNA templates in G_1- and S-phase reactions. This suggested that the deficiency in G_1-promoted replication was in the initiation process. Both in prokaryotic cell-free replication systems and in the SV40 system, one of the earliest events in replication of duplex DNA is site-specific binding of an initiator protein to the replication origin, followed by unwinding of the origin DNA by a helicase (Stahl et al. 1986; Dean et al. 1987; Dodson et al. 1987; Wold et al. 1987; Kornberg 1988). In the SV40 system, both origin binding and the helicase activity are properties of T antigen. However, in addition to T antigen, cellular factors are also required for efficient DNA unwinding at the SV40 replication origin (Wold et al. 1987). To demonstrate directly that initiation was limiting in G_1 extracts, we have used origin unwinding as an assay for the initiation process. We asked whether the increased activity of S-phase extracts was due to an increased ability to promote site-specific unwinding of DNA at the replication origin.

T-antigen-mediated DNA unwinding was assayed by incubating together a plasmid containing the SV40 replication origin, cell extract, T antigen, ATP, and topoisomerase I at 37°C for 2 hours (Fig. 2). During the reaction, the template DNA was immediately relaxed by topoisomerase I. Then, as the covalently closed, circular DNA was unwound by the T-antigen helicase, the compensating positive supercoils induced in the template were removed by the topoisomerase. In this reaction, the cell extract supplied a single-stranded DNA-binding protein that stabilized the unwound DNA, promoting extensive unwinding by T antigen. Following deproteinization, as the product becomes fully base paired over the formerly unwound region, the DNA must become negatively supercoiled, with the number of negative supercoils dependent on the number of bases unwound by the T-antigen helicase. The unwound (supercoiled) products were visualized on agarose gels. The unwinding reaction was performed on radiolabeled DNA, permitting direct quantitation of the extent of the reaction.

Increasing amounts of G_1- or S-phase extract were tested for their ability to promote T-antigen-mediated unwinding of the SV40 replication origin (Fig. 2). The amounts of extract used were those that gave a linear increase in the amount of total DNA replication in our standard replication reaction. The difference in the ability of G_1-phase versus S-phase extracts to promote origin unwinding duplicated the difference almost precisely in these extracts to support DNA replication. S-phase extracts were tenfold more effective at promoting origin unwinding than G_1 extracts were. If origin unwinding, as detected here, was a necessary step in the formation of an active replication complex, then this difference was sufficient to

A

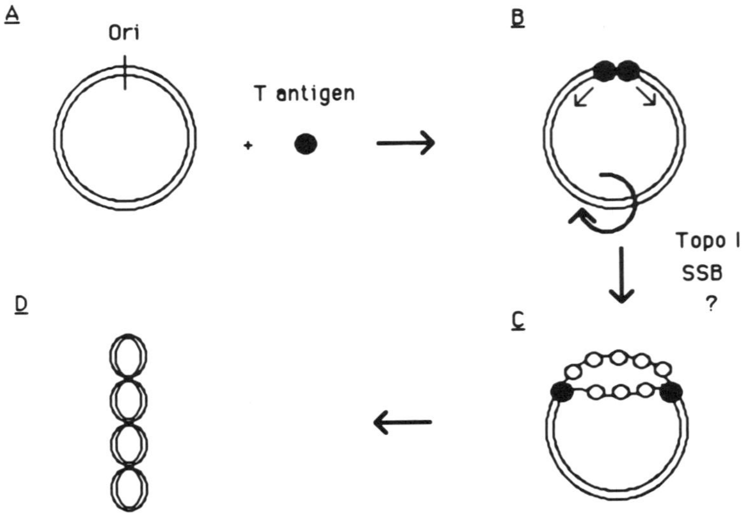

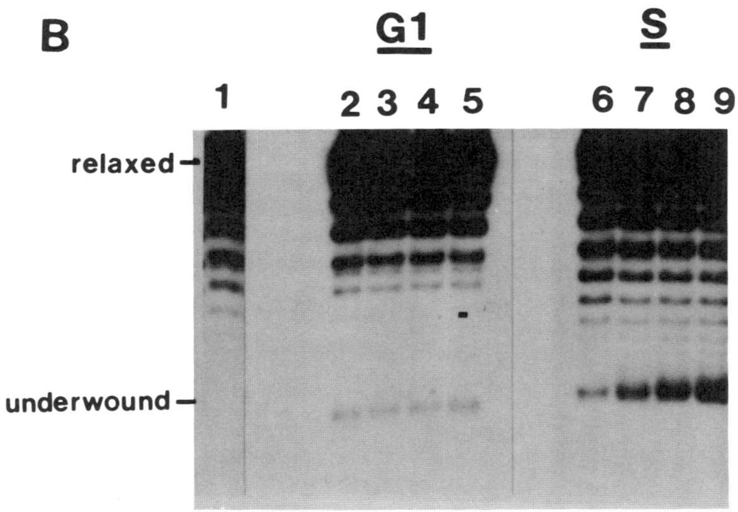

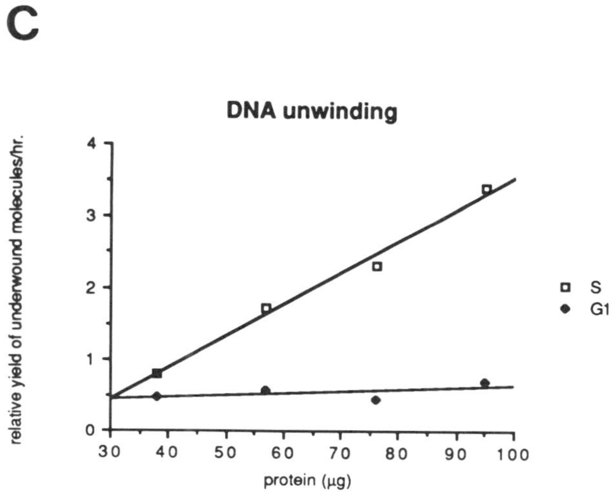

Figure 2 T-antigen-dependent origin unwinding. (*A*) A schematic diagram for the pathway of T-antigen-dependent origin unwinding. This reaction is described in detail in the text. (*B*) DNA unwinding in G_1 and S-phase extracts. The unwinding reactions were performed as described in the text. In this experiment, the substrate DNA was labeled with ^{32}P, permitting the reaction products to be visualized by autoradiography (see text). Note that SV40-origin-dependent unwinding requires that the reactions be performed in at least 50 mM NaCl. (Lane *1*) Control reaction containing T antigen but no cytoplasmic extract; (lanes *2–5*) T antigen plus 38, 57, 76, or 95 µg of G_1 extract, respectively; (lanes *6–9*) T antigen plus 38, 57, 76, or 95 µg of S-phase extract, respectively. Control reactions (not shown) demonstrate that the unwinding reaction is T-antigen-dependent, but the addition of exogenous topoisomerase is not necessary, consistent with our observation that the S-100 extracts contain 0.25 units of topoisomerase per microgram of protein. Although the addition of exogenous topoisomerase I is not necessary, we always include it in our reactions to ensure that it is present in excess. (*C*) The region of the gel shown in panel *B* containing the underwound product was cut out, and the relative yield of product was calculated by measuring the amount of ^{32}P label present in each band.

explain the increased capacity of S-phase extracts to replicate DNA.

Although it is an early event, an unknown number of steps precede origin DNA unwinding during initiation. Therefore, the increase in origin unwinding detected in S-phase extracts might be a consequence of activation of a more upstream event. Purification of the S-phase activator, by complementation of G_1 replication activity, will be necessary before we can address this issue definitively.

CONCLUSION

We have begun to study the molecular events associated with the cell's decision to begin DNA replication. To this end, we developed a cell-free DNA replication assay, on the basis of SV40 replication, that faithfully reproduces some of the cell-cycle-related changes in the cell's capacity to replicate DNA. One interpretation of our data leads to the following scenario for the onset of DNA replication during the cell cycle. Most factors necessary to replicate DNA are equally abundant at all phases of the cell cycle. However, the efficient use of these factors to replicate DNA requires that they have functional or physical access to a replication origin. Site-specific unwinding of DNA at replication origins, in providing that access, may be the limiting step in DNA replication. One consequence of the cell's decision to begin DNA replication would be synthesis of a factor(s) that is necessary for DNA unwinding at replication origins. This factor is synthesized at the G_1 boundary. Furthermore, we speculate that the mechanism that limits DNA replication in eukaryotic cells to one round per cell cycle will involve a *cis*-acting process that limits the origin unwinding to just once per cell cycle.

REFERENCES

Baserga, R. 1984. Growth in size and cell DNA replication. *Exp. Cell Res.* **151:** 1.

Dean, F., P. Bullock, Y. Murakami, R. Wobbe, L. Weissbach, and J. Hurwitz. 1987. Simian virus 40 DNA replication: SV40 large T antigen unwinds DNA containing the SV40 origin of replication. *Proc. Natl. Acad. Sci.* **84:** 16.

Dodson, M., F. Dean, P. Bullock, H. Echols, and J. Hurwitz. 1987. Unwinding of duplex DNA from the SV40 origin of replication by T antigen. *Science* **238:** 964.

Kornberg, A. 1988. DNA replication. *J. Biol. Chem.* **263:** 1.

Li, J. and T. Kelly. 1984. Simian virus 40 DNA replication in vitro. *Proc. Natl. Acad. Sci.* **81:** 6973.

Stahl, H. and R. Knippers. 1983. SV40 large T antigen on replicating viral chromatin: Tight binding and localization on the viral genome. *J. Virol.* **47:** 65.

Stahl, H., P. Droge, and R. Knippers. 1986. DNA helicase activity of SV40 large tumor antigen. *EMBO J.* **5:** 1939.

Stillman, B. and Y. Gluzman. 1985. Replication and supercoiling of SV40 DNA in cell extracts from human cells. *Mol. Cell. Biol.* **5:** 2051.

Thompson, C., P. Challoner, P. Neiman, and M. Groudine. 1985. Levels of c-*myc* oncogene mRNA are invariant throughout the cell cycle. *Nature* **314:** 363.

———. 1986. Expression of the c-*myb* proto-oncogene during cellular proliferation. *Nature* **319:** 374.

Wold, M., J. Li, and T. Kelly. 1987. Initiation of simian virus 40 DNA replication in vitro: Large tumor antigen and origin dependent unwinding of the template. *Proc. Natl. Acad. Sci.* **84:** 3643.

Mammalian Cell Mutants Temperature Sensitive in DNA Synthesis

J.J. Dermody, K.G. Lawlor, L. Malkas, H. Du, M. Orlian, B. Wojcik, K.K. Jha, and H.L. Ozer

Department of Biological Sciences, Hunter College
City University of New York, New York, 10021

Temperature-sensitive (*ts*) mutants have been a powerful tool in understanding the molecular mechanism of many cellular and viral processes. However, only limited studies have been described for mammalian cells in culture, and few mutants have been defined as defective in functions directly related to DNA synthesis (*ts*DNA), in contrast to those in cell-cycle progression. Indeed, the greatest insights into mammalian DNA synthesis have come from studies involving replication of DNA viruses and their temperature-sensitive mutants intracellularly and, more recently, in vitro (see Stillman et al., this volume). Our laboratory has sought to combine the dual approaches of cell genetics and virus model systems in an effort to define novel cellular functions in DNA synthesis. Several years ago, this laboratory described the isolation of temperature-sensitive mutants in 3T3 that affected cellular and polyomavirus DNA synthesis (Slater and Ozer 1976; Zeng et al. 1984). These mutants (*ts*2, *ts*20) mapped to the X chromosome and complemented poorly, or not at all, each other and several other mouse *ts*DNA mutants, despite evident phenotypic differences and all being recessive to wild type (Jha et al. 1980; Dermody et al. 1988). More recently, we shifted our attention to the Chinese hamster cell line (CHO) as the parent for a mutant collection on the basis of its favorable growth properties, amenability to replica plating, and efficiency in expressing recessive mutations. We speculated that mutants at autosomal loci might be easier to obtain in these cells than in other cell lines. Additionally, CHO cells have been found to replicate efficiently the genomes of two well-studied but distinct classes of DNA viruses: human adenoviruses (Ad2 and Ad5) (Longiaru and Horwitz 1981; Radna et al. 1987) and mouse polyomavirus (LaBella and Ozer 1985).

RESULTS

We have developed a strategy for obtaining mammalian cell mutants temperature sensitive for DNA synthesis, on the basis of the considerations mentioned previously. First, we have isolated over 200 mutants in CHO (and another Chinese hamster cell line, V79) that are temperature sensitive for growth by utilizing a variety of selective and nonselective strategies, including replica plating (Ozer et al. 1987; Dermody et al. 1986). Approximately 10% were temperature sensitive for DNA synthesis. Second, each of the *ts*DNA mutants were evaluated for Ad2 and polyomavirus DNA synthesis at temperatures permissive (33°C) and restrictive (39°C) for cellular DNA synthesis. All of the nine Chinese hamster mutants examined supported viral DNA synthesis at 33°C. Five of the mutants restricted polyomavirus DNA synthesis in a temperature-dependent manner. Adenovirus DNA synthesis was restricted in only one of these mutants (JB3-B) and in neither of the 3T3 mutants studied previously (*ts*2, *ts*20). The conditions of infection made it highly unlikely that a defect in cell-cycle progression was responsible; the differential effect on adenovirus and polyomavirus DNA synthesis ruled out a defect in nucleotide pool metabolism. On the basis of these and other data, the latter is indeed probably responsible for the *ts*DNA phenotype in JB3-B. Results obtained with these and other *ts*DNA mutants obtained from other laboratories are summarized in Table 1.

Table 1 Viral DNA Replication in *ts*DNA Mutants

Mutant	Cell line	Adenovirus	Polyoma
JB1-C	CHO	+	+
JB1-R	CHO	+	+
JB3-B	CHO	–	–
JB3-O	CHO	+	–
JB6-N	CHO	+	+
JB7-K	CHO	+	–
JB8-D	CHO	+	–
JB11-J	CHO	+	–
JB17-D	V79	+	+
*ts*C8	CHO	+	+
*ts*13A	CHO	+	+
*ts*2	3T3	+	–
*ts*20	3T3	+	–
*ts*A1S9	L	+	+
*ts*C1	L	+	–

Analysis involving pair-wise cell fusion of the CHO *ts*DNA mutants indicates that multiple complementation groups are represented.

As a prelude to determining biochemically the function that is heat-labile, we have adapted the in vitro system for DNA synthesis initially developed for SV40 by Li and Kelly (1984). Cell extracts prepared from CHO cells supplemented with immunoaffinity-purified T antigen support the replication of a plasmid containing origin sequences derived from the homologous papovavirus DNA. In the case of SV40, the efficiency of synthesis is rather low and markedly reduced, as compared to extracts prepared from permissive monkey cells. When the same cell extracts were utilized for polyomavirus DNA synthesis, the level of replication with the "polyoma system" was considerably higher than with the "SV40 system." Polyomavirus DNA synthesis is higher in extracts from both mouse L cells and 3T3 cells (subline A31N) than in CHO extracts, attaining rates consistent with that reported by Murakami et al. (1986) for the mouse cell line FM3A and comparable to that of SV40 in its permissive system. In all cases, polyomavirus DNA replication was reduced at 39°C in extracts prepared even from wild-type cells.

We have begun the analysis of the temperature-sensitive cell mutants by examining *ts*20 in vitro. This mutant was isolated from A31N and has been shown to restrict intracellular polyomavirus DNA synthesis at an early stage of strand elongation (Zeng et al. 1984). We have found that viral DNA synthesis is reduced in *ts*20 in a temperature-independent fashion, although the defect is even more marked at temperatures above 33°C. Enzyme purification studies performed in collaboration with the laboratory of E. Baril (Vishwanatha et al. 1986; Dermody et al. 1988) provide a plausible basis for this result, as *ts*20 has been found to have a defect in a purified multiprotein form of DNA polymerase α. Only 20–25% of the activity is found as polymerase α2, whether DNA polymerase is purified from cells grown at 33°C or cells shifted to the restrictive temperature for 16 hours. In contrast, wild-type cells and a non-temperature-sensitive revertant (*ts*20R) show greater than 90% of the activity chromatographing as polymerase α2 at either temperature, demonstrating that the property is linked genetically to the temperature-sensitive mutation.

Other aspects of the *ts*20 phenotype, however, suggest that the defect may be more complicated. First, a defect in topoisomerase I has also been observed (Zeng et al. 1984, 1985). In addition, studies in intact cells with *ts*2, another 3T3 mutant that does not complement *ts*20, suggest that the function encoded by the mutation may have its primary effect at a stage prior to DNA synthesis itself. Density-arrested *ts*2 cells can be induced to resume cell growth and DNA synthesis at 33°C by serum stimulation. As expected, cultures shifted to 39°C after serum addition do not undergo DNA synthesis. However, they also fail to show the expected increase in thymidine kinase (TK) and other enzymes associated with DNA synthesis, including thymidine monophosphate kinase (TMPK) and DNA polymerase α. The defect is at the level of mRNA metabolism because such cultures also fail to accumulate TK mRNA.

DISCUSSION

We have isolated a large collection of temperature-sensitive mutants in rodent cells and identified a small subset that restricts viral and cellular DNA synthesis in a temperature-dependent fashion. We anticipate that such mutants will be valuable in defining further functions essential to DNA replication. We have shown that the in vitro systems developed for the DNA papovaviruses, SV40 and polyomavirus, in their conventional permissive cells (primate and mouse, respectively) can be extended to the CHO cell system in which most of our *ts*DNA mutants have been isolated. Analysis of polyomavirus DNA synthesis in such extracts may permit in vitro complementation to facilitate identification of the basis for the temperature-sensitive cell defect.

Because not all of these *ts*DNA mutants would be expected to be defective in currently devised cell-free systems, we have also begun efforts to isolate the gene responsible for the temperature-sensitive defect by DNA-mediated gene transfer utilizing a human cosmid library (Lau and Kan 1983). We have successfully obtained primary transfectants for five of the mutants at a frequency consistent with the temperature-sensitive defect being corrected by a unique copy sequence. Because the human sequences were functional in a cosmid library, the intact gene in each case would be expected to be smaller than 40–45 kb. Human-specific DNA sequences have been recovered from transfectants of *ts*20. We are currently determining whether this sequence is responsible for correcting the temperature-sensitive defect in *ts*20. Such cloned sequences should also provide the basis for assessment of the genetic basis for the noncomplementing mouse mutants, including *ts*2.

ACKNOWLEDGMENTS

This work was supported by grants from the National Institutes of Health to H.L.O. (CA-23002) and to the Research Centers in Minority Institutions, Center for Gene Structure and Function at Hunter College (RR0307).

REFERENCES

Dermody, J.J., B.E. Wojcik, H. Du, and H.L. Ozer. 1986. Identification of temperature-sensitive DNA⁻ mutants of Chinese hamster cells affected in cellular and viral DNA synthesis. *Mol. Cell. Biol.* **6:** 4594.

Dermody, J.J., K.G. Lawlor, H. Du, B. Wojcik, K.K. Jha, L. Malkas, R. Hickey, E.F. Baril, and H.L. Ozer. 1988. Polyoma virus DNA synthesis *in vitro:* Studies with CHO, 3T3, and their tsDNA mutants. *Cancer Cells* **6:** 95.

Jha, K.K., M. Siniscalco, and H.L. Ozer. 1980. Temperature-sensitive mutants of the BALB/3T3 cells. III. Hybrids between ts2 and other mouse mutant cells affected in DNA synthesis and correction of ts2 defect by human X chromosome. *Somatic Cell Mol. Genet.* **6:** 603.

La Bella, F. and H.L. Ozer. 1985. Differential replication of SV40 and polyoma DNA in Chinese hamster ovary cells. *Virus Res.* **2:** 329.

Lau, Y.-F. and Y.W. Kan. 1983. Versatile cosmid vectors for the isolation, expression, and rescue of gene sequences: Studies with the human β-globin gene cluster. *Proc. Natl. Acad. Sci.* **80:** 5225.

Li, J.J. and T.J. Kelly. 1984. Simian virus 40 DNA replicates *in vitro. Proc. Natl. Acad. Sci.* **81:** 6973.

Longiaru, M. and M.S. Horwitz. 1981. Chinese hamster ovary cells replicate adenovirus DNA. *Mol. Cell. Biol.* **1:** 208.

Murakami, Y., T. Eki, M. Yamada, C. Prives, and J. Hurwitz. 1986. Species-specific in vitro synthesis of DNA containing the polyoma virus origin of replication. *Proc. Natl. Acad. Sci.* **83:** 6347.

Ozer, H.L., J.J. Dermody, and B. Wojcik. 1987. Mammalian ts DNA mutants. *UCLA Symp. Mol. Cell. Biol. New Ser.* **47:** 475.

Radna, R., B. Foellmer, L.A. Feldman, U. Francke, and H.L. Ozer. 1987. Restriction of human adenovirus replication in Chinese hamster cell lines and their hybrids with human cells. *Virus Res.* **8:** 277.

Slater, M. and H.L. Ozer. 1976. Temperature-sensitive mutants of BALB/3T3 cells. II. Description of a mutant affected in cellular and polyoma virus DNA synthesis. *Cell* **7:** 289.

Vishwanatha, J.K., S.A. Coughlin, M. Wesolowski-Owen, and E.F. Baril. 1986. A multiprotein form of DNA polymerase from HeLa cells: Resolution of its associated catalytic activities. *J. Biol. Chem.* **261:** 6619.

Zeng, G.-C., J. Donegan, H.L. Ozer, and R. Hand. 1984. Characterization of a ts mutant of BALB/3T3 cells and correction of the defect by in vitro addition of extracts from wild-type cells. *Mol. Cell. Biol.* **4:** 1815.

Zeng, G.-C., M. Zannis-Hadjopoulos, H.L. Ozer, and R. Hand. 1985. Defective DNA topoisomerase I activity in a DNA ts mutant of BALB/3T3 cells. *Somatic Cell Mol. Genet.* **11:** 557.

Regulated DNA Replication and Nuclear Assembly in Cell-free Systems from *Xenopus* Eggs

R.A. Laskey, J.J. Blow, M.A. Sheehan, S.M. Dilworth, and A.D. Mills

Cancer Research Campaign Molecular Embryology Research Group
Department of Zoology, University of Cambridge
Cambridge, CB2 3EJ, England

The study of eukaryotic DNA replication has been restrained by the lack of systems that initiate nonviral replication in vitro. This lack of systems has prevented fractionation to identify regulatory factors. Eggs of the frog *Xenopus laevis* provide a unique opportunity to study initiation of nonviral replication in vitro, because it is now possible to prepare egg homogenates in ways that support efficient initiation and completion of DNA replication.

Xenopus eggs have allowed the development of cell-free systems that assemble DNA into nucleosomes (Laskey et al. 1977; Dilworth et al. 1987), decondense demembraned sperm nuclei, assemble nuclear envelopes around them (Lohka and Masui 1983, 1984), and assemble purified DNA into pseudonuclei surrounded by functional nuclear envelopes (Blow and Laskey 1986; Newmeyer et al. 1986a,b; Newport 1987; Sheehan et al. 1988). In addition, they initiate semiconservative DNA replication in vitro, using either demembraned nuclei or purified DNA as templates (Blow and Laskey 1986, 1988; Blow and Watson 1987; Hutchinson et al. 1987; Newport 1987). The development of these systems has been made possible by the extraordinary maternal stockpiles of materials required for chromosome replication and nuclear assembly (Laskey et al. 1985). These accumulate during oocyte growth to allow the *Xenopus* embryo to complete consecutive cell cycles in only 30 minutes.

This paper surveys progress in understanding the process of nuclear assembly and the regulation of DNA replication using cell-free systems from *Xenopus* eggs. Synthetic and assembly capacities of extracts prepared in different ways are summarized in Table 1.

Table 1 Replication and Nuclear Assembly in High- or Low-speed Supernatants of Activated or Unactivated *Xenopus* Eggs

	Unactivated egg supernatant		Activated egg supernatant	
	10,000*g*	100,000*g*	10,000*g*	100,000*g*
Nucleosome assembly	+	+	+	+
Nuclear envelope assembly	–	–	+	–
Nuclear envelope disassembly and chromosome condensation	+	+	–	–
DNA synthesis on single-stranded templates	+	+	+	+
Semiconservative DNA replication on duplex templates	–	–	+	–

Steps in the Assembly of Replication-competent Nuclei

When purified DNA is added to high- or low-speed supernatants of *Xenopus* eggs, it is rapidly assembled into regularly spaced chains of nucleosomes. Each mononucleate egg can assemble DNA equivalent to 10,000 diploid nuclei in approximately 1 hour. This process is mediated by two acidic proteins. Nucleoplasmin binds and transfers histones H2A and H2B to DNA (Laskey et al. 1978; Dilworth et al. 1987), and N1 binds and transfers histones H3 and H4 (Kleinschmidt et al. 1984; Dilworth et al. 1987).

It is not yet clear if new histones are assembled onto nuclei added to the extracts as well as onto purified DNA, but nuclei undergo extensive decondensation. When low-speed supernatants of activated eggs are used, nuclear envelopes form around the decondensing chromatin (Lohka and Masui 1983, 1984). However, nuclear envelopes do not form in 100,000*g* supernatants. Instead, a membrane fraction is required. It has been possible to manipulate nuclear envelope assembly and structure by varying the ratio of supernatant and membrane fractions. This has revealed that efficient formation of immature nuclear pores can precede formation of detectable nuclear membranes but that both soluble and membrane fractions are required for pore assembly (Sheehan et al. 1988). It remains possible that very small islands of membrane are required to nucleate pore formation. When nuclear structure is manipulated in this way, DNA replication is observed only after nuclear envelope formation (Sheehan et al. 1988).

Initiation of Replication In Vitro on Nuclei or Purified DNA

Lohka and Masui (1983) observed that decondensed sperm nuclei incorporated [^{3}H]thymidine during incubation in low-speed supernatants of activated eggs. Blow and Laskey (1986) showed that this represents efficient semiconservative replication. Furthermore, Blow and Watson (1987) showed that *Xenopus* sperm nuclei replicate completely by flow cytometry of fluorescently labeled nuclei. Their DNA content increases from haploid to diploid in approximately 1 hour in the cell-free system. At least 70% of nuclei can be induced to replicate in the egg extract. It is striking that replication is constrained to only one round in a single cell cycle (see below).

Purified DNA also replicates after incubation in low-speed supernatants (Blow and Laskey 1986; Newport 1987), resembling its behavior after injection into intact eggs (Harland and Laskey 1980). Interestingly, purified DNA is assembled into pseudonuclei after injection into intact eggs (Forbes et al. 1983) or after incubation in vitro (Blow and Laskey 1986; Newmeyer et al. 1986a,b; Newport 1987). Correlations have been observed between replication efficiency and nuclear assembly (Blow and Laskey 1986; Sheehan et al. 1988), and efficient replication has not yet been reported in the absence of nuclear assembly.

Nuclei Act as Integrated and Independent Units of Replication within a Single Incubation

The replication of individual nuclei within an incubation has been analyzed by flow cytometry, following incorporation of biotinylated dUTP into DNA and probing with fluorescent streptavidin (Blow and Watson 1987). In this way, nuclei were observed to replicate completely, but asynchronously, so that some nuclei had doubled their DNA content before others had started to replicate. Furthermore, the fastest nuclei completed replication in only 30 minutes, resembling in vitro rates and indicating that initiations must be coordinated and near synchronous throughout an individual nucleus, even though they were asynchronous with respect to other nuclei.

Replication Is Restained to Once per Cell Cycle without a Requirement for Specific DNA Sequences

The most striking feature of the cell-free replication system derived from *Xenopus* eggs is its precise regulation of replication, allowing only one round in a single cell cycle (Blow and Laskey 1986; Blow and Watson 1987). This is seen most clearly when the template are nuclei, rather than purified DNA; but provided care is taken to avoid mechanical damage of fragile pseudonuclei (see below), it is also true of purified DNA templates. This mimics results obtained when DNA is microinjected into intact unfertilized eggs (Harland and Laskey 1980; Mechali et al. 1983). However, these experiments have produced another more surprising result, namely, that the mechanism that prevents reinitiation of replication within a single cell cycle does not require specific DNA sequences. Although these experiments do not allow the conclusion that specific replication origins are not used in other types of cells or even in the endogenous chromosome of *Xenopus* eggs, they do show clearly that replication under strict cell-cycle control occurs for a range of entirely prokaryotic templates (Harland and Laskey 1980; Mechali et al. 1983; Laskey et al. 1985). Thus, bacteriophage λ or G4, as well as many other templates, are able to replicate multiple times in eggs when repeated cell cycles are allowed, but only once during a single cell cycle or when the cell-cycle clock is arrested. Thus, in *Xenopus* eggs, there can be no requirement for specific *cis*-acting DNA sequences for the mechanism that prevents reinitiation of replication within a single cell cycle. This contrasts with the mechanism that restricts bovine papillomavirus to a single round of replication per cell cycle (Berg et al. 1986; Roberts and Weintraub 1986, 1988).

A Role for the Nuclear Envelope in Regulating DNA Replication during the Cell Cycle

Previously, we have discussed possible models for restricting DNA replication to one round per cell cycle (Laskey et al. 1981; Blow et al. 1987). Our recent attempts to test these models have asked which feature of passage through mitosis is necessary to allow a replicated nucleus to replicate again. The results of these experiments are described elsewhere (Blow and Laskey 1988). They show that the only mitotic change that is necessary to allow a second cycle of replication is permeabilization of the nuclear envelope. Thus, when nuclei are replicated in the absence of protein synthesis, they are unable to rereplicate when transferred to fresh extract. However, if replicated nuclei are treated with lysolecithin—a lecithin analog that inserts into membranes permeabilizing them—and are then transferred to fresh extract, a second round of replication is observed. Other mechanical or biochemical methods of permeabilizing the nuclear envelope also allow a second round of replication without undergoing chromosome condensation.

These observations suggest a single model for the coupling of DNA replication to the cell cycle (Blow and Laskey 1988), in which an essential initiation factor lacks a nuclear migration signal and is unable to cross the nuclear envelope. Thus, it could only gain access to DNA when the nuclear envelope breaks down at mitosis. It would then bind to DNA to allow initiation of replication after envelope assembly is complete. It would be inactivated during replication, but no new factor could gain access to the DNA to allow a second round of replication until the nuclear envelope breaks down again at the next mitosis. Further aspects of this model are described by Blow and Laskey (1988).

ACKNOWLEDGMENTS

We are grateful to the Cancer Research Campaign for generous support. J.J.B. is a Beit Junior Research Fellow.

REFERENCES

Berg, L., M. Lusky, A. Stenlund, and M.R. Botchan. 1986. Repression of bovine papilloma virus replication is mediated by a virally encoded *trans*-acting factor. *Cell* **46:** 753.

Blow, J.J. and R.A. Laskey. 1986. Initiation of DNA replication in nuclei and purified DNA by a cell-free extract of *Xenopus* eggs. *Cell* **47:** 557.

———. 1988. A role for the nuclear envelope in controlling DNA replication within the cell cycle. *Nature* **332:** 546.

Blow, J.J. and J.V. Watson. 1987. Nuclei act as independent and integrated units of replication in a *Xenopus* cell-free replication system. *EMBO J.* **6:** 1997.

Blow, J.J., S.M. Dilworth, A.D. Mills, and R.A. Laskey. 1987. Chromosome replication in a cell-free system from *Xenopus* eggs. *Philos. Trans. R. Soc. Lond. B* **317:** 483.

Dilworth, S.M., S.J. Black, and R.A. Laskey. 1987. Two complexes that contain histones are required for nucleosome assembly *in vitro*: Role of nucleoplasmin and N1 in *Xenopus* egg extracts. *Cell* **51:** 1009.

Forbes, D.J., M.W. Kirschner, and J.W. Newport. 1983. Spontaneous formation of nucleus-like structures around bacteriophage DNA microinjected into *Xenopus* eggs. *Cell* **44:** 23.

Harland, R.M. and R.A. Laskey. 1980. Regulated replication of DNA microinjected into eggs of *Xenopus laevis*. *Cell* **21:** 761.

Hutchison, C.J., R. Cox, R.S. Drepaul, M. Gomperts, and C.C. Ford. 1987. Periodic DNA synthesis in cell-free extracts of *Xenopus* eggs. *EMBO J.* **6:** 2003.

Kleinschmidt, J.A., E. Fortkamp, G. Krohne, H. Zentgraf, and W.W. Franke. 1984. Co-existence of two different types of soluble histone complexes in nuclei of *Xenopus laevis* oocytes. *J. Biol. Chem.* **260:** 1166.

Laskey, R.A., S.E. Kearsey, and M. Mechali. 1985. Analysis of chromosome replication with eggs of *Xenopus laevis*. In *Genetic engineering: Principles and methods* (ed. J.K. Setlow and A. Hollaender), vol. 7, p. 135. Plenum Press, New York.

Laskey, R.A., A.D. Mills, and N.R. Morris. 1977. Assembly of SV40 chromatin in a cell-free system from *Xenopus* eggs. *Cell* **10:** 237.

Laskey, R.A., R.M. Harland, W.C. Earnshaw, and C. Dingwall. 1981. Chromatin assembly and the co-ordination of DNA replication in the eukaryotic genome. In *International cell biology 1980-1981* (ed. H.G. Schwieger), p. 161. Springer-Verlag, Berlin.

Laskey, R.A., B.M. Honda, A.D. Mills, and J.T. Finch. 1987. Nucleosomes are assembled by an acidic protein which binds histones and transfers them to DNA. *Nature* **275:** 416.

Lohka, M.J. and Y. Masui. 1983. Formation *in vitro* of sperm pronuclei and mitotic chromosomes by amphibian ooplasmic components. *Science* **220:** 719.

———. 1984. Roles of cytosol and cytoplasmic particles in nuclear envelope assembly and sperm pronuclear formation in cell-free preparations from amphibian eggs. *J. Cell. Biol.* **98:** 1222.

Mechali, M., F. Mechali, and R.A. Laskey. 1983. Tumor promoter TPA increases initiation of replication on DNA injected into *Xenopus* eggs. *Cell* **35:** 63.

Newmeyer, D.D., D.R. Finlay, and D.J. Forbes. 1986a. *In vitro* transport of a fluorescent nuclear protein and exclusion of non-nuclear proteins. *J. Cell Biol.* **103:** 2091.

Newmeyer, D.D., J.M. Lucocq, T.R. Burglin, and E.M. De Robertis. 1986b. Assembly *in vitro* of nuclei active in nuclear protein transport: ATP is required for nucleoplasmin accumulation. *EMBO J.* **5:** 501.

Newport, J. 1987. Nuclear reconstitution *in vitro*: Stages of assembly around protein-free DNA. *Cell* **48:** 205.

Roberts, J.M. and H. Weintraub. 1986. Negative control of DNA replication in composite SV40-bovine papilloma virus plasmids. *Cell* **46:** 741.

———. 1988. Cis-acting negative control of DNA replication in eukaryotic cells. *Cell* **52:** 397.

Sheehan, M.A., A.D. Mills, A.M. Sleeman, R.A. Laskey, and J.J. Blow. 1988. Steps in the assembly of replication-competent nuclei in a cell-free system from *Xenopus* eggs. *J. Cell. Biol.* **106:** 1.

Human Homolog of the Yeast *cdc2+/CDC28* Cell-cycle Regulator: Cyclic Activation of the Protein Kinase during Mitosis

G. Draetta and D. Beach

Cold Spring Harbor Laboratory, Cold Spring Harbor, New York 11724

The *cdc2*$^+$ gene of *Schizosaccharomyces pombe* and its homolog in *Saccharomyces cerevisiae, CDC28,* play a fundamental role in the control of cell division in yeast (Beach et al. 1981; Reed and Nasmyth 1981; Reed et al., this volume). Each gene encodes a protein kinase that shares 62% amino acid identity (Lorincz and Reed 1984; Hindley and Phear 1984; Simanis and Nurse 1986) and whose biochemical activities appear to be functionally interchangeable, because either gene can rescue temperature-sensitive defects of the other (Beach et al. 1981; Booher and Beach 1986). Using antibodies raised against the yeast *cdc2*$^+$ gene product, a human homolog (p34) of this protein kinase has been identified (Draetta et al. 1987). Like the yeast proteins, human p34 displays protein kinase activity in immunoprecipitates, and like the *cdc2*$^+$ gene product, it has been found to interact with a 13-kD protein that is homologous to the *suc1*$^+$ gene product of *S. pombe* (Draetta et al. 1987). Furthermore, the human gene encoding p34 has been isolated by rescue of a temperature-sensitive *cdc2* mutant of fission yeast (Lee and Nurse 1987).

In a recent study (Draetta et al. 1988), we have shown that synthesis of p34 in baby rat kidney cells, following infection with adenovirus, reflects the proliferative state of the cell. This result is compatible with the possibility of involvement of p34 in regulation of the cell cycle in mammalian cells but provides no direct evidence of a specific role.

To investigate the properties of p34 during the cell cycle, we have made use of HeLa cells grown in suspension culture. These cells are amenable to the technique of synchronization by elutriation centrifugation. The following properties of HeLa cell p34 have been demonstrated (see Draetta and Beach 1988).

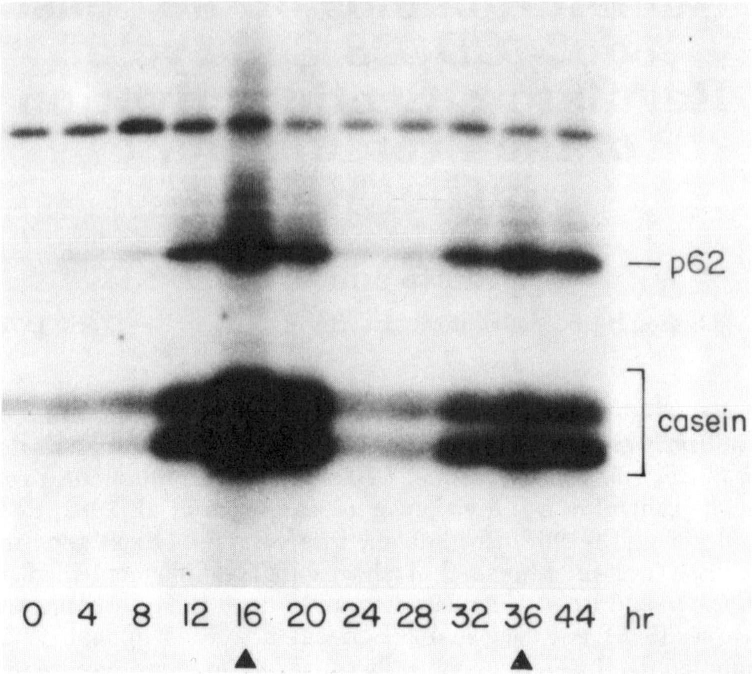

Figure 1 Periodic activation of p34 protein kinase. G_1-enriched cells obtained by elutriation were inoculated into fresh medium and sampled at 4-hr intervals. Protein kinase activity in antipeptide precipitates was assayed in the presence of caseine. Solid triangles mark the times at which the highest proportion of cells were in G_2/M, determined by flow cytometry.

(1) In vivo, p34 is phosphorylated at two or more sites; (2) the protein exists in a multiprotein complex with the p13^{suc1} homolog and also with a newly identified polypeptide of 62 kD (p62); (3) p62 is found in a complex with only the most heavily phosphorylated form of p34; (4) p34 phosphorylates both the endogenous substrate p62 and also casein and (5) various properties of p34 are subject to cell-cycle regulation. Using centrifugal elutriation, we have found that while the p34 levels are constant throughout the cycle, its state of phosphorylation, its association with p62, and the kinase activity of the p34/p62 complex are selectively increased in G_2/M-enriched cell fractions.

The cyclic pattern of activation of p34 kinase in a synchronous culture of HeLa cells is illustrated in Figure 1. The kinase is abruptly activated as cells enter mitosis and is deactivated rapidly thereafter. We have also found that in cells ex-

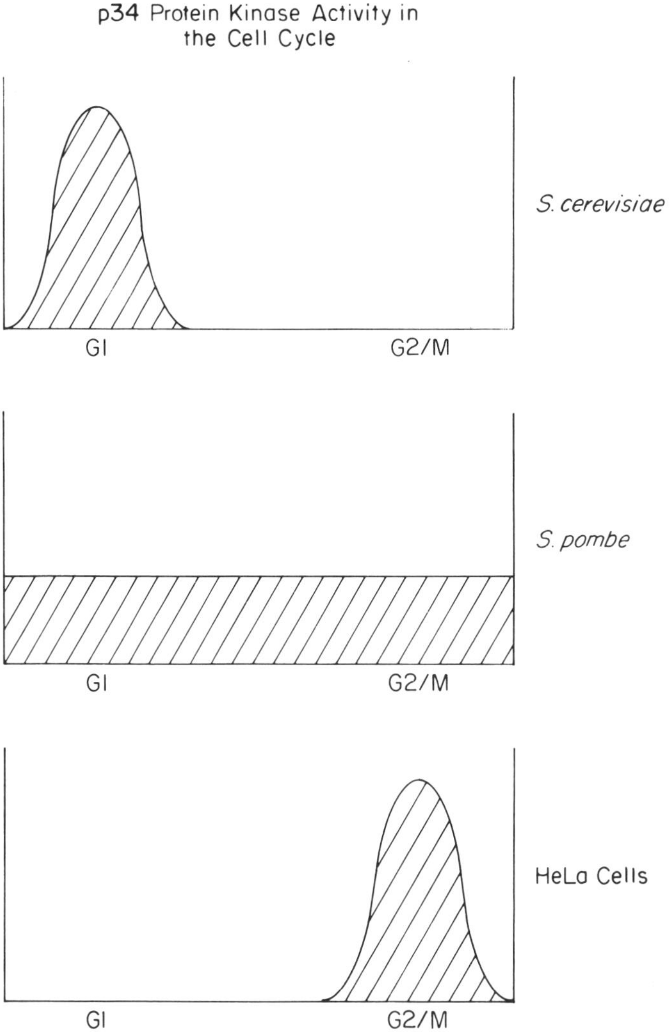

Figure 2 Differential activation of the *S. cerevisiae* p36^{CDC28}, *S. pombe* p34^{cdc2}, and HeLa cell p34 protein kinases during the cell cycle.

posed to nocodazole, an agent that causes mitotic arrest by disrupting microtubules (De Brabander et al. 1976), the p34 kinase accumulates in a fully activated state. This observation confirms that p34 activation is selectively occurring during mitosis.

These experiments raise several interesting questions about the role of p34 protein kinase in different organisms. Figure 2 summarizes the presently known cell-cycle properties of p36^{CDC28}, p34^{cdc2}, and p34 from *S. cerevisiae, S. pombe*, and HeLa cells, respectively. In *S. cerevisiae*, the phosphorylation of a 40-kD protein, which acts as a substrate in immunoprecipitates of p36^{CDC28}, is enhanced during G_1 and is inactive during mitosis (Mendenhall et al. 1987). p36^{CDC28} does not seem to phosphorylate any tested exogenous artificial substrates efficiently, and thus some ambiguity exists in assessing alterations in its activity. However, p34 from fission yeast will phosphorylate casein in vitro. With respect to this substrate, the activity of the protein kinase does not fluctuate during the cell cycle (Simanis and Nurse 1986; L. Brizuela and D. Beach, unpubl.). Both of these observations contrast with the behavior of p34 in HeLa cells, in which the protein kinase is dramatically activated during mitotic metaphase.

The biochemical properties of p34 during the cell cycle have to be interpreted in the light of genetic observations that have defined the role of this protein kinase in the yeast cell cycle. In fission yeast, *cdc2*$^+$ is required both for the initiation of nuclear DNA replication and independently for the transition from G_2 to mitosis (Nurse and Bissett 1981). In budding yeast, the G_1 role of *CDC28* has been clearly established (Hartwell 1974). It also appears that the gene is required for mitotic initiation, but in *S. cerevisiae*, this step occurs early in the cell cycle, shortly after commitment to DNA synthesis (see Piggott et al. 1982). Thus, activation of the protein kinase during the G_1/S transition is not inconsistent with the role in both DNA replication and mitotic control.

In this context, the activation of p34 protein kinase exclusively during mitosis in HeLa cells is a slight puzzle. These observations are clearly compatible with the kinase playing a role in mitosis but do not conspicuously point to a role in G_1. Further investigation of this issue will require application of techniques that inhibit the activation of p34 in mammalian cells.

The observations described here do not define the role of human p34 precisely. However, in an accompanying paper (Dunphy et al., this volume), p34 is shown to play a critical role as a mitotic activator in *Xenopus* oocytes. This role is fully consistent with the biochemical properties of p34 in human cells.

REFERENCES

Beach, D.H., B. Durkacz, and P.M. Nurse. 1981. Functionally homologous cell cycle control genes in budding and fission yeast. *Nature* **300:** 706.

Booher, R. and D.H. Beach. 1986. Site-specific mutagenesis of *cdc2*$^+$ a cell cycle control gene of the fission yeast *Schizosaccharomyces pombe*. *Mol. Cell. Biol.* **6:** 3523.

Brizuela, L., G. Draetta, and D. Beach. 1987. p13^{suc1} acts in the fission yeast cell division cycle as a component of the p34^{cdc2} protein kinase. *EMBO J.* **6:** 3507.

De Brabander, M.J., R.M.L. Van de Veire, F.E.M. Aerts, M. Borgers, and P.A.J. Janssen. 1976. The effects of methyl [5-(2-thienylcarbonyl)-1H-benzimidazol-2-yl-carbamate, (R 17934; NSC 238159), a new synthetic antitumoral drug interfering with microtubules, on mammalian cells cultured *in vitro*. *Cancer Res.* **36:** 905.

Draetta, G. and D. Beach. 1988. Activation of the *cdc2* protein kinase during mitosis in human cells: Cell-cycle dependent phosphorylation and subunit rearrangement. *Cell* (in press).

Draetta, G., D. Beach, and E. Moran. 1988. Synthesis of p34, the mammalian homologue of the yeast *cdc2*$^+$*/CDC28* protein kinase, is stimulated during adenovirus-induced proliferation of baby rat kidney cells. *Oncogene* (in press).

Draetta, G., L. Brizuela, J. Potashkin, and D. Beach. 1987. Identification of p34 and p13, human homologs of the cell cycle regulators in fission yeast encoded by *cdc2*$^+$ and *suc1*$^+$. *Cell* **50:** 319.

Hartwell, L.H. 1974. *Saccharomyces cerevisiae* cell cycle. *Bacteriol. Rev.* **38:** 164.

Hindley, J. and G.A. Phear. 1984. Sequence of the cell division gene *cdc2*$^+$ from *Schizosaccharomyces pombe*, patterns of splicing and homology to protein kinases. *Gene* **31:** 129.

Lee, M. and P. Nurse. 1987. Complementation used to clone a human homologue of the fission yeast cell cycle control gene *cdc2*$^+$. *Nature* **327:** 31.

Lorincz, A.T. and S.I. Reed. 1984. Primary structure homology between the product of yeast cell division control gene *CDC28* and vertebrate oncogenes. *Nature* **307:** 183.

Mendenhall, M.D., C.A. Jones, and S.I. Reed. 1987. Dual regulation of the yeast CDC28-p40 protein kinase complex: Cell cycle, pheromone and nutrient limitation effects. *Cell* **50:** 927.

Nurse, P. and Y. Bissett. 1981. Gene required in G_1 for commitment to cell cycle and in G_2 for control of mitosis in fission yeast. *Nature* **292:** 558.

Piggott, J.A., R. Rai, and B.L.A. Carter. 1982. A bifunctional gene product involved in two phases of the cell cycle. *Nature* **298:** 391.

Reed, S. and K.A. Nasmyth. 1981. Isolation of genes by complementation in yeast: Molecular cloning of a cell cycle gene. *Proc. Natl. Acad. Sci.* **77:** 2119.

Simanis, V. and P. Nurse. 1986. The cell cycle control gene *cdc2*$^+$ of yeast encodes a protein kinase potentially by phosphorylation. *Cell* **45:** 261.

A Yeast *cdc* Gene Product Regulates Mitotic Conversion in *Xenopus* Egg Extracts

W.G. Dunphy,[1] L. Brizuela,[2] D. Beach,[2] and J. Newport[1]

[1]Department of Biology, University of California
San Diego, La Jolla, California 92093

[2]Cold Spring Harbor Laboratory, Cold Spring Harbor
New York 11724

During the conversion to the mitotic state, higher eukaryotic cells activate a cascade of reactions, resulting in the disintegration of nuclear envelope, the condensation of the DNA into chromosomes, and the reorganization of the cytoskeleton. In *Xenopus*, the induction of the mitotic state appears to be under the control of a cytoplasmic factor known as maturation-promoting factor, or MPF (Masui and Markert 1971; Gerhart et al. 1984). This mitosis-inducing factor has been detected in diverse species, ranging from human to yeast. The mechanism of action of MPF is a key problem in cell-cycle regulation. In fission yeast, there has been considerable progress in the genetic elucidation of regulatory networks controlling the G_2 to M transition (see Nurse 1985). The relationship of the presently identified gene products to MPF has been unclear.

Cell-free Assay for MPF

By utilizing recent technical advances in the construction and disassembly of nuclei in vitro (Forbes et al. 1983; Lohka and Masui 1983; Newport and Spann 1987), we devised a rapid and sensitive version of an in vitro assay for MPF (Dunphy and Newport 1988). The assay employs reconstituted nuclei in interphase cytoplasm from activated *Xenopus* eggs. The MPF-induced conversion from interphase to mitosis is monitored conveniently by the visual observation of the loss of the nuclear envelope from the substrate nuclei. Crude MPF preparations can induce nuclear breakdown even at 50-fold dilutions. This system is therefore approximately tenfold more sensitive than the traditional oocyte-maturation assay (Gerhart et al. 1984) and previous cell-free systems (Lohka and Maller 1985; Miake-

Lye and Kirschner 1985). At maximal concentrations of MPF, nuclear breakdown requires 20–30 minutes. Preincubation experiments have revealed that the action of MPF requires only a few minutes and that the disassembly process itself takes up the remainder of the incubation period. Because the cell-free extracts are stable to freezing and thawing, the same preparation can be used continuously over several months.

Fate of MPF during Interphase

Using the cell-free system, we set out to follow the fate of MPF between mitotic periods. The inactivation of MPF upon the entry into interphase could occur by two mechanisms. MPF could be either irreversibly destroyed or reversibly modified. We reasoned that if the latter were the case, it might be possible to reactivate the putatively latent MPF in interphase cytoplasm. Indeed, we found that the addition of as little as 2.5% by weight of active MPF reactivated a latent reservoir of MPF in interphase cytoplasm. The regenerated levels of MPF were quite similar to those of unfertilized eggs, which contain their entire complement of MPF in an active form. These experiments suggest that there are two components in mitotic control. One is MPF, a constitutive factor that executes mitotic rearrangements. The other is an activator of MPF, which must be synthesized recurrently in each cell cycle. The requirement for this component explains the necessity for protein synthesis in each cell cycle. The finding that MPF is present constantly throughout the cell cycle eliminates the possibility that MPF itself is a cyclin (Evans et al. 1983). The activator of MPF would be expected to behave as a cyclin-like molecule.

The *Schizosaccharomyces pombe suc1* Gene Product Inhibits Mitotic Conversion

As an ongoing interest, we have attempted to take advantage of the known yeast *cdc* mutants to aid in the molecular characterization of the MPF-induced conversion to the mitotic state in vitro. These attempts have recently led to an exciting breakthrough. We have found that the 13-kD product of the *Schizosaccharomyces pombe suc1* gene (Hayles et al. 1986; Hindley et al. 1986; Brizuela et al. 1987;) is a potent and specific inhibitor of MPF-induced nuclear disassembly. *suc1* was isolated (Hayles et al. 1986) as a suppressor of the *cdc2* gene (Nurse 1985). Maximal inhibition occurs at 2.5 μM p13 (~30 μg protein/ml). This represents about 0.1% of the total protein in the in vitro incubation. Even at concentrations 50-fold higher, p13 has no effect on the intricate process of in vitro nuclear assembly around sperm chromatin. The inhibition by p13 is reversible. Finally, the inhibitory activity is destroyed by trypsin but is unaffected by boiling. This thermostability is characteristic of many low-molecular-weight polypeptides.

Xenopus cdc2 Homologs Bind to a p13 Affinity Column

We set out to elucidate the target(s) of inhibition by p13 and the stage in MPF-induced disassembly at which the inhibition occurred. Because p13 is active at micromolar concentrations, we reasoned that affinity chromatography with p13-agarose might be useful for this purpose. From genetic and biochemical studies in *S. pombe* (Hayles et al. 1986; Brizuela et al. 1987), p13 is known to interact with the product of the *cdc2* gene, a 34K protein. We have found that *Xenopus* eggs contain two proteins (34K and 33K) that cross-react with polyclonal antisera against the yeast *cdc2* protein. When mitotic extracts from *Xenopus* eggs were applied to p13-agarose, both of the *Xenopus cdc2* homologs were quantitatively adsorbed to the column. No adsorption to mock-coupled agarose occurred. These polypeptides bound to p13-agarose very tightly. Only strong denaturants were effective in elution of the bound proteins from the column. These experiments demonstrate that yeast p13 can physically associate with *Xenopus cdc2* homologs. The results do not prove, however, that this association is responsible for the mitotic inhibition that we observe.

p13 Affinity Chromatography Depletes MPF

In principle, p13 could inhibit MPF-induced nuclear disassembly at any of three stages: it could interfere directly with MPF, it could inhibit transduction of the response to MPF, or it could block the process of nuclear disassembly. Our recent work suggests that p13 interferes with MPF itself, perhaps by direct physical association. When mitotic extracts or crude MPF preparations were chromatographed on p13-agarose, the MPF activity was quantitatively depleted from the extract. No depletion occurred during chromatography on mock-coupled agarose or on agarose containing any of six irrelevant (control) polypeptides. We have not been able to elute active MPF from the affinity column as yet.

Barring nonspecific inactivation of MPF by the affinity column, there are at least three models to account for the retention of MPF on p13-agarose: First, either or both of the *Xenopus cdc2* homologs might constitute MPF. This simple model seems unlikely, since the bulk of the *cdc2* homologs do not cofractionate with active MPF during several chromatographic steps. It is more likely that active MPF is a complex of another polypeptide(s) and the *cdc2* protein. Finally, it is conceivable that p13 may bind directly to protein(s) other than the *cdc2* products.

DISCUSSION

Two overall strategies have been employed for elucidation of the regulatory mechanisms that govern the progression of the cell cycle. In genetically tractable organisms such as budding and fission yeast (Nurse 1985), a collection of mutants with perturbations in the regulation of the cell cycle has been generated. On the other hand, in the *Xenopus* egg/oocyte system, one can effectively examine several steps of the mitotic process (e.g., nuclear disassembly and reassembly) in cell-free reactions. As we have demonstrated here, the product of a yeast gene (*suc1*) can function in a *Xenopus* in vitro system to control the entry into mitosis. Therefore, it is now possible to combine the respective advantages of the yeast and frog systems for unraveling the mechanisms of mitotic control.

REFERENCES

Brizuela, L., G. Draetta, and D. Beach. 1987. $p13^{suc1}$ acts in fission yeast cell division cycle as a component of the $p34^{cdc2}$ protein kinase. *EMBO J.* **6:** 3507.

Dunphy, W. and J. Newport. 1988. Mitosis-inducing factors are present in a latent form during interphase in the *Xenopus* embryo. *J. Cell Biol.* (in press).

Evans, T., E. Rosenthal, J. Youngblom, D. Distel, and T. Hunt. 1983. Cyclin, a protein specified by maternal mRNA in sea urchin eggs that is destroyed at each cleavage division. *Cell* **33:** 389.

Forbes, D., M. Kirschner, and J. Newport. 1983. Spontaneous formation of nucleus-like structures around bacteriophage DNA microinjected into *Xenopus* eggs. *Cell* **34:** 13.

Gerhart, J., M. Wu, and M. Kirschner. 1984. Cell cycle dynamics of an M-phase-specific cytoplasmic factor in *Xenopus laevis* oocytes and eggs. *J. Cell Biol.* **98:** 1247.

Hayles, J., D. Beach, B. Durkacz, and P. Nurse. 1986. The fission yeast cell cycle control gene *cdc2*; isolation of a sequence *suc1* that suppresses *cdc2* mutant function. *Mol. Gen. Genet.* **202:** 291.

Hindley, J., G. Phear, M. Stein, and D. Beach. 1987. *Suc1*⁺ encodes a predicted 13-kilodalton protein that is essential for cell viability and is directly involved in the division cycle of *Schizosaccharomyces pombe. Mol. Cell. Biol.* **7:** 504.

Lohka, M. and J. Maller. 1985. Induction of nuclear envelope breakdown, chromosome condensation, and spindle formation in cell-free extracts. *J. Cell Biol.* **101:** 518.

Lohka, M. and Y. Masui. 1983. Formation *in vitro* of sperm pronuclei and mitotic chromosomes induced by amphibian ooplasmic components. *Science* **220:** 719.

Masui, Y. and C. Markert. 1971. Cytoplasmic control of nuclear behavior during meiotic maturation of frog oocytes. *J. Exp. Zool.* **177:** 129.

Miake-Lye, R. and M. Kirschner. 1985. Induction of early mitotic events in a cell-free system. *Cell* **41:** 165.

Newport, J. and T. Spann. 1987. Disassembly of the nucleus in mitotic extracts: Membrane vesicularization, lamin disassembly, and chromosome condensation are independent processes. *Cell* **48:** 219.

Nurse, P. 1985. Cell cycle control genes in yeast. *Trends Genet.* **1:** 51.

Properties of Purified Maturation-promoting Factor from *Xenopus* Eggs

J.L. Maller, E. Erikson, and M.J. Lohka

Department of Pharmacology
University of Colorado School of Medicine
Denver, Colorado 80262

The regulation of the eukaryotic cell cycle has fundamental significance for the control of cell proliferation and the process of morphogenesis and differentiation in early development. Two general restriction points in the cell cycle have been identified by genetic and biochemical analyses. One control point is at the G_0/G_1 transition, and its regulation by peptide growth factors and oncogenes has been well documented (Pardee et al. 1978; Marcus et al. 1985). The other control point is at the G_2/M border, where cell-cycle progression can be blocked by mutations in certain genes that encode protein kinase activities (Reed et al. 1985; Russell and Nurse 1987). In other cases, arrest at the G_2/M border occurs when the level of cAMP is elevated (Nose and Katsuta 1975; Stambrook and Velez 1976).

Evidence both from cell-fusion studies with mitotic and interphase cells and from injection of interphase cells with mitotic cytoplasm has shown that mitosis (M phase) is dominant over interphase (Masui and Markert 1971; Rao 1982). The cytoplasmic M-phase activity that drives cells into mitosis is known as maturation-promoting factor (MPF), and its characterization has represented a major unresolved problem in cell-cycle control. Recently, our laboratory developed a cell-free system from *Xenopus* eggs that could assemble nuclei in vitro in a manner that could then respond to MPF by undergoing nuclear envelope breakdown, chromosome condensation, and spindle formation (Lohka and Maller 1985). This breakthrough in inducing early mitotic events in vitro formed the basis for a new assay for MPF that was markedly superior to the previous assay, which utilized microinjection of *Xenopus* oocytes. The advantages of the cell-free assay include the ability to detect an intermediate response, the ability to detect activity in diluted fractions, which is difficult in the microinjection assay (Wu and

Gerhart 1980), and the capability of increasing the concentration of MPF in the assay system to 67% of the total volume versus 10% in the injection assay. We have used breakdown of pronuclei in the cell-free system to purify MPF 3500-fold to near homogeneity. Purified MPF consists of two proteins with relative molecular weights of 45,000 and 32,000, which are able to induce germinal vesicle breakdown when microinjected into *Xenopus* oocytes (Lohka et al. 1988). Purified MPF expresses a protein kinase activity directed toward the 45,000-M_r protein and can also phosphorylate histone H1, phosphatase inhibitor-1, and casein. MPF is unable to phosphorylate a number of other commonly used in vitro substrates, such as glycogen synthase, ribosomal protein S6, lamins, and vimentin.

Several experimental questions about MPF are now readily amenable to biochemical analysis. One question is, Why is β-glycerophosphate required for detection of MPF activity in vivo and in vitro with both crude and purified MPFs (Tachibana et al. 1987; Lohka et al. 1988)? This question is still unresolved, but the MPF protein kinase activity is inhibited 50% by 100 mM β-glycerophosphate, ruling out a stimulatory role of this compound on MPF-directed phosphorylation. Another question concerns how MPF is activated. It is known that MPF is activated posttranslationally in oocytes injected with catalytic amounts of active MPF (Reynhout and Smith 1974), and recently, this activation of stored oocyte latent MPF has been achieved in a cell-free system (Cyert and Kirschner 1988). When MPF is activated, there is a large increase in oocyte protein phosphorylation (Maller et al. 1977), and in vitro activation of latent oocyte MPF requires ATP (Cyert and Kirschner 1988), suggesting that MPF is activated by phosphorylation. Recently, we have obtained direct evidence for this hypothesis by demonstrating that purified MPF kinase activity can be deactivated by dephosphorylation by protein phosphatase 2A. Readdition of more MPF to dephosphorylated MPF does not restore MPF activity, suggesting that some other element is involved in phosphorylating and activating MPF; i.e., activation of MPF is not autocatalytic. This result is in contrast to previous results, where crude preparations of MPF were able to cause activation of latent MPF (Reynhout and Smith 1974; Cyert and Kirschner 1988). The nature of this other element is currently unknown but could be hypothesized to be a cyclin or a cdc protein because these often encode protein kinase activities involved in the G_2/M transition and, in some cases, can activate MPF upon injection into oocytes (Swenson et al. 1986).

The identification of this putative MPF-activating element may be related to the ability of γ-S-ATP to stabilize MPF during extraction from eggs. γ-S-ATP affords a means to accumulate proteins irreversibly in their phosphorylated forms because γ-S-ATP can be used by many protein kinases for thiophosphorylation, but protein phosphatases generally cannot dephosphorylate thiophosphorylated proteins (Gratecos and Fischer 1974). During our purification of MPF, we found it absolutely essential to include γ-S-ATP during the extraction for activity to be subsequently detectable. In early stages of purification, further addition of γ-S-ATP to MPF increases its activity as much as fourfold, but at higher degrees of purification, γ-S-ATP has no effect on MPF activity itself (S. Yamashita and J. Maller, unpubl.). In addition, incubation of purified MPF with [γ-^{35}S]ATP does not label either the 45,000- or 32,000-M_r protein, indicating that MPF itself is not thiophosphorylated. Recently, we found that purified MPF could reactivate MPF from interphase cytosols, but it was essential to include γ-S-ATP for reactivation. These results indicate that the target of thiophosphorylation is not MPF itself but, rather, some other protein needed to activate MPF, which is stabilized by thiophosphorylation. However, present data cannot rule out the converse hypothesis, namely, that the MPF-inactivating system is permanently inactivated by thiophosphorylation.

A final experimental question that we are addressing concerns how the general increase in protein phosphorylation is achieved by MPF. Although MPF itself appears to be a protein kinase and thus must have interesting and important substrates besides those already identified, some proteins such as ribosomal protein S6 and lamins undergo identical cell-cycle oscillations in phosphate content in concert with MPF activity but are not substrates in vitro for MPF kinase themselves. To address the mechanism of MPF action, we have chosen to study the pathway by which MPF activates S6 kinase. Because S6 kinase can also phosphorylate lamins, this pathway may also be involved in the direct stimulation of nuclear envelope breakdown via lamin depolymerization after dephosphorylation. S6 kinase II that is activated during oocyte maturation has been purified to homogeneity from unfertilized eggs (Erikson and Maller 1986), antibodies have been made to it (Erikson et al.

1987), and the gene has been cloned and sequenced (Jones et al. 1988). The purified S6 kinase II can be deactivated by serine phosphatases, but no reactivation or phosphorylation of the S6 kinase occurs with MPF kinase, even though S6 kinase II is activated immediately in oocytes injected with MPF. Phosphorylation and reactivation of the deactivated S6 kinase is evident with a protein kinase from 3T3 cells highly specific for MAP-2 (T.W. Sturgill et al., in prep.). This MAP kinase can also be deactivated by dephosphorylation, but MPF kinase is unable to reactivate it, indicating that there must be at least one additional intermediate between MPF and S6 kinase II. Further analysis of this pathway should reconstitute the entire sequence of activating signals by which MPF activates S6 kinase II, and presumably this will form a general model for how MPF causes pleiotropic changes in phosphorylation via activation of numerous protein kinases.

ACKNOWLEDGMENT

This work is supported by grants from the National Institutes of Health (GM-26743 and DK-28353) and the American Cancer Society (NP-517C and CD-279).

REFERENCES

Cyert, M.S. and M. Kirschner. 1988. Regulation of MPF activity in vitro. *Cell* **53:** 185,

Erikson, E. and J.L. Maller. 1986. Purification and characterization of a protein kinase from *Xenopus* eggs highly specific for ribosomal protein S6. *J. Biol. Chem.* **261:** 350.

Erikson, E., D. Stefanovic, J. Blenis, R.L. Erikson, and J.L. Maller. 1987. Antibodies to *Xenopus* egg S6 kinase II recognize S6 kinase from progesterone- and insulin-stimulated oocytes and from proliferating chicken embryo fibroblasts. *Mol. Cell. Biol.* **7:** 3147.

Gratecos, D. and E.H. Fischer. 1974. Adenosine 5′-0(3-thiotriphosphate) in the control of phosphorylase activity. *Biochem. Biophys. Res. Commun.* **58:** 960.

Jones, S.W., E. Erikson, J. Blenis, J. Maller, and R.L. Erikson. 1988. A *Xenopus* ribosomal protein S6 kinase has two apparent kinase domains that are each similar to distinct protein kinases. *Proc. Natl. Acad. Sci.* **85:** 3377.

Lohka, M.J. and J.L. Maller. 1985. Induction of nuclear envelope breakdown chromosome condensation and spindle formation in cell-free extracts. *J. Cell. Biol.* **101:** 518.

Lohka, M.J., M.K. Hayes, and J.L. Maller. 1988. Purification of maturation- promoting factor, an intracellular regulator of early mitotic events. *Proc. Natl. Acad. Sci.* **85:** 3009.

Maller, J.L., M. Wu, and J.C. Gerhart. 1977. Changes in protein phosphorylation accompanying maturation of *Xenopus laevis* oocytes. *Dev. Biol.* **58:** 295.

Marcus, M., A. Fainsod, and G. Diamond. 1985. The genetic analysis of mammalian cell-cycle mutants. *Annu. Rev. Genet.* **19:** 389.

Masui, Y. and C.L. Markert. 1971. Cytoplasmic control of nuclear behavior during meiotic maturation of frog oocytes. *J. Exp. Zool.* **177:** 129.

Nose, K. and H. Katsuta. 1975. Arrest of cultured rat liver cells in G_2 phase by the treatment with dibutyryl cAMP. *Biochem. Biophys. Res. Commun.* **64:** 983.

Pardee, A.B., R. Dubrow, J.L. Hamlin, and R.F. Kletzien. 1978. Animal cell cycle. *Annu. Rev. Biochem.* **47:** 715.

Rao, P.N. 1982. The phenomenon of premature chromosome condensation. In *Premature chromosome condensation* (ed. P.N. Rao et al.), p. 1. Academic Press, New York.

Reed, S.I., J.A. Hadwiger, and A.T. Lorincz. 1985. Protein kinase activity associated with the product of the yeast cell division cycle gene *CDC28*. *Proc. Natl. Acad. Sci.* **82:** 4055.

Reynhout, J.K. and L.D. Smith. 1974. Studies on the appearance and nature of a maturation-inducing factor in the cytoplasm of amphibian oocytes exposed to progesterone. *Dev. Biol.* **38:** 394.

Russell, P. and P. Nurse. 1987. The mitotic inducer *nim*1$^+$ functions in a regulatory network of protein kinase homologs controlling the initiation of mitosis. *Cell* **49:** 569.

Stambrook, P. and C. Velez. 1976. Reversible arrest of Chinese hamster V79 cells in G_2 by dibutyryl cyclic AMP. *Exp. Cell Res.* **99:** 57.

Swenson, K.I., K.M. Farrell, and J.V Ruderman. 1986. The clam embryo protein cyclin A induces entry into M-phase and the resumption of meiosis in *Xenopus* oocytes. *Cell* **47:** 861.

Tachibana, K., N. Yangagishima, and T. Kishimoto. 1987. Preliminary characterization of maturation-promoting factor from yeast *Saccaromyces cerevisiae*. *J. Cell. Sci.* **88:** 273.

Wu, M. and J.C. Gerhart. 1980. Partial purification and characterization of the maturation-promoting factor from eggs of *Xenopus laevis*. *Dev. Biol.* **79:** 465.

Protein Synthesis, Proteolysis, and the Control of Cell Division in Early Embryos: Do the Synthesis and Destruction of Cyclin Comprise the Cytoplasmic Oscillator?

J. Minshull,[1] J. Pines,[1] N. Standart,[1] L. Stewart,[1,2] S. Mackie,[1] A. Colman,[3] J. Blow,[4] M. Wu,[5] J.V. Ruderman,[6] and T. Hunt[1]

[1]Department of Biochemistry, University of Cambridge
Cambridge, CB2 1QW, England

[3]Department of Biochemistry, University of Birmingham
Birmingham, England

[4]Department of Zoology, University of Cambridge
Cambridge, England

[5]Department of Molecular Biology, University of California
Berkeley, California 94720

[6]Department of Zoology, Duke University, Durham
North Carolina 27706

The cyclins are a family of proteins that were first recognized in fertilized sea urchin eggs and clam oocytes (Evans et al. 1983; Swenson et al. 1986) and subsequently were found in other marine invertebrate eggs (Standart et al. 1987). Cyclin mRNA is present in frog oocytes and eggs (J. Minshull, unpubl.). The isolation of a *Drosophila* cyclin was reported by P. O'Farrell (O'Farrell et al., this volume), and a more distantly related yeast (*Saccharomyces cerevisiae*) homolog was identified on the basis of sequence homology by S. Reed (Reed et al., this volume). In eggs, cyclins are encoded by abundant "masked" maternal mRNAs, which are not translated to a significant extent before fertilization but load rapidly onto polysomes after fertilization. The cyclin signature is their sud-

[2]Present address: Oncogene Science, Inc., Manhasset, New York 11030.

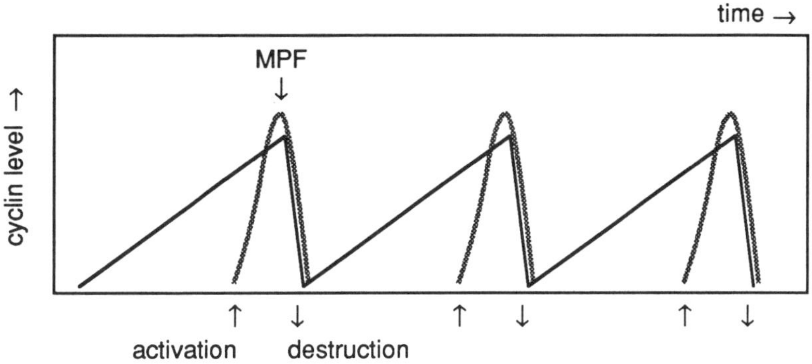

Figure 1 Cyclin levels show a sawtooth oscillation, whereas MPF levels rise more abruptly. This diagram summarizes current understanding of the relationship between cyclin and MPF. Cyclin is produced by translation of its mRNA, so its levels rise smoothly throughout the cell cycle. MPF activity appears abruptly at the transition from G_2 to M. MPF activity falls abruptly when cyclin is destroyed. The following questions arise: What is the nature of MPF activation? Does it simply require cyclin levels to exceed a critical threshold, or does it represent a response to an additional signal? What triggers and what performs the destruction of cyclin? Does this destruction kill MPF activity? If so, how?

den disappearance every time the cell divides. This disappearance is thought to be due to proteolysis. Studies of the kinetics of polypeptide synthesis and destruction in the first two mitotic cell cycles show that cyclin is stable during interphase but is destroyed in the space of 5 minutes or less just before the chromosomes part at the metaphase-anaphase transition of the cell cycle. If cells are prevented from entering mitosis, cyclin is as stable as any other protein specified by maternal mRNA, lasting for many hours. Such is the case in clam or sea urchin eggs when DNA synthesis is inhibited by hydroxyurea or aphidicolin.

Because the concentration of cyclin follows a simple sawtooth oscillation in step with cleavage (see Fig. 1), it is a natural candidate for the role of catalyzing entry into mitosis in cleaving eggs. Many studies show that protein synthesis is required in each cell cycle to enter M phase in cleaving sea urchin eggs (Hultin 1961; Wilt et al. 1967; Wagenaar 1983). The protein (or proteins) required for division is specified by maternal mRNA, because inhibition of RNA synthesis does not block entry into M phase. New protein synthesis is required continuously, so the proteins needed for division must be consumed during the cell cycle. Either they are unstable or they are assembled into stable structures that accumulate during cell division (e.g., chromosomes, centrosomes, and kinetochores). The periodic instability of the cyclins corresponds exactly to the first of these expectations. Obviously, other proteins are involved in the control of cell division. The cyclins must interact with other cellular components, most of which are formed during oogenesis. Doubtless other proteins encoded by maternal mRNA belong in the category of those that form stable complexes during early development. Histone mRNAs, for example, are frequently stored in unfertilized eggs (Woodland 1982).

Cyclin Destruction Requires a Bipolar Spindle: Colchicine Holds Cells in Mitosis by Slowing Cyclin's Proteolysis

The first evidence that destruction of cyclin might be necessary for passage through the cell cycle came from studies of colchicine-treated and parthenogenetically activated clam and sea urchin eggs. Cyclin destruction is retarded and impaired in the presence of colchicine; the level of cyclin thus remains high at the time when it would normally drop precipitously. The same kind of effect occurs in slightly less extreme form in parthenogenetic clam embryos, which have intact microtubules but form monopolar spindles (Kuriyama et al. 1986). Such cells remain in mitosis for several hours instead of a few minutes. However, when protein synthesis is inhibited in parthenogenetically activated clam oocytes, or colchicine-arrested clam and sea urchin eggs, cyclin levels do eventually drop to zero. The cells reenter interphase at precisely the point when cyclin finally disappears (T. Hunt and J. Ruderman, unpubl.). Controls in which protein synthesis is not inhibited stay in mitosis much longer and, in some cases, almost indefinitely. It is not clear how cells with monopolar or nonexistent spindles ever escape from their mitotic trap, but both KCl-activated oocytes and colchicine-treated cells that are allowed to continue protein synthesis eventually exit from mitosis (albeit highly asynchronously). They pass rapidly through interphase, spend a long time in M phase, and so on, without ever dividing.

Protease Inhibitors Stabilize Cyclin and Block Starfish Oocytes in M Phase

The protease inhibitors leupeptin, soybean trypsin inhibitor, and antipain block 1-methyladenine-treated starfish oocytes in first meiotic metaphase, and MPF (maturation-promoting factor) levels rise normally but stay high (Picard et al. 1985). Antipain prevents cyclin destruction (the other two inhibitors have not been tested) (J. Pines et al., unpubl.), and we presume that this accounts for the meiotic arrest. It would be very helpful to know how to inhibit cyclin destruction specifically, but first meiosis in starfish oocytes is the only case where we have had success. The identification and characterization of the destruction mechanism is a high priority. Whatever the protease is, it is highly specific, for cyclins appear to be the only targets among proteins specified by maternal mRNA. The protease shows astonishing regulation, for cyclin is only unstable during mitosis and then only briefly during the moments preceding the metaphase-anaphase transition. It is surely significant that microtubule integrity is necessary for the destruction, but the significance of this observation in terms of a possible mechanism is unclear.

Clam and Sea Urchin Cyclin mRNA Make Frog Oocytes Enter M Phase

The first experiment supporting a direct role for cyclin in promoting the transition from G_2 to M was performed by Swenson et al. (1986), who showed that injection of synthetic clam cyclin A mRNA into *Xenopus* oocytes promoted their maturation into eggs. This experiment has been repeated with sea urchin maternal mRNA and synthetic cyclin mRNA (Pines and Hunt 1987). The newly synthesized sea urchin cyclin protein is clearly visible as a labeled gel band, and it disappears from the *Xenopus* oocytes at about the time of white spot formation, presumably at the end of meiosis I metaphase. Thus, frog oocytes can both respond to and destroy a heterologous cyclin. As little as 0.3 ng of "synthetic" cyclin mRNA, the lowest dose tested, still gave 100% maturation, and we estimate that 0.01 ng of natural mRNA should be sufficient. This makes cyclin mRNA one of the most potent artificial inducers of oocyte maturation.

We tested whether cyclin mRNA was the active principle in total sea urchin maternal mRNA that makes frog oocytes mature (it is possible that others might exist) by removing its cyclin mRNA with antisense deoxynucleotides and RNase H before injection into the frog oocytes (Minshull and Hunt 1986). This treatment abolished the MPF-like activity of the mRNA. These results suggest that cyclin mRNA is sufficient and necessary to promote passage from G_2 to M in stage-6 *Xenopus* oocytes. The question remains, however, whether this result is physiologically meaningful or whether it is another example of the ever-growing list of interesting agents that are active in this assay, like insulin, cAMP-dependent protein kinase inhibitors, and *ras* oncogenes (Maller and Krebs 1980; Birchmeier et al. 1985; Maller 1985).

Xenopus Oocytes Contain (at Least) Two Different Cyclin mRNAs

The natural signal for *Xenopus* oocyte maturation is the steroid hormone, progesterone. Progesterone-induced maturation requires protein synthesis (but not RNA synthesis), suggesting that the hormone either stimulates the translation of maternal mRNA or stabilizes a rapidly synthesized but unstable protein. The protein(s) in question presumably activates the endogenous pool of pre-MPF. We decided to search for such mRNAs in frog oocytes and eggs, suspecting strongly that we might find *Xenopus* cyclin.

We repeated the total mRNA injections that had been successful in the case of sea urchin maternal mRNA, except that *Xenopus* eggs or oocytes provided the poly(A)$^+$ RNA. We found that *Xenopus* egg poly(A)$^+$ RNA was quite active in promoting the maturation of *Xenopus* oocytes, whereas oocyte poly(A)$^+$ RNA was only about 20% as active. Conversely, oocyte poly(Asa)$^-$ RNA showed activity, whereas egg poly(A)$^-$ RNA did not. These results suggest that an mRNA encoding an MPF-like activity is present in nonadenylated form in oocytes and that this mRNA lengthens its poly(A) tail during meiotic maturation. Such a class of mRNAs is commonly found in eggs and oocytes (Rosenthal et al. 1983; Dworkin et al. 1985).

We looked for cyclin mRNA in frogs by making a library from *Sau*3A-cut cDNA from *Xenopus* egg poly(A)$^+$ RNA inserted into *Bam*HI-cut M13mp8. This library was screened in parallel with two probes: a consensus oligonucleotide, based on a conserved region between clam and urchin cyclins, and an urchin cDNA clone, corresponding to the complete coding region of the mRNA. A positive clone was identified and sequenced. It showed strong homology with clam and urchin cyclin. The same library was subsequently rescreened with the oligonucleotide alone, and about 20 independent new clones were sequenced. These clones all corresponded to a second type of cyclin mRNA, which shows less homology with sea urchin cyclin. Both cyclins were found as full-length coding region clones in λgt10 cDNA libraries. They correspond to mRNAs of roughly equal abundance, with about 10^7 molecules of each message per oocyte. Variant frog cyclin 1 and frog cyclin 2 clones differing from one another mainly at third base positions in codons exist. A typical frog thus has at least four types of cyclin mRNAs, denoted 1A, 1B, 2A, and 2B. The extent of variation in these mRNAs between individual frogs is not known.

Cutting Cyclin mRNA with Antisense Oligodeoxyribonucleotides Sometimes Blocks Maturation and Always Blocks Cleavage

In determining the best way to test whether cyclin was necessary for cell division, it seemed that some form of HART (hybrid arrest of translation) might be successful once we knew the sequence(s) of cyclin mRNAs. We thought that short oligodeoxyribonucleotides might be useful, because it would be possible to synthesize them chemically and because relatively high concentrations could be used to drive rapid annealing. In addition, small molecules would probably diffuse rapidly through the cytoplasm, and DNA was likely to be more stable than RNA. Once the importance of RNase H had been appreciated, it was simple to show that polysomal globin mRNA could be cut to completion rapidly, even while it was being translated, by oligonucleotides containing only 10–12 nucleotides (Minshull and Hunt 1986; Dash et al. 1987; Shuttleworth and Colman 1988). When anticyclin oligonucleotides were injected into *Xenopus* oocytes, we found that cyclin mRNA was cut completely in about 30 minutes.

The first oligonucleotide we used (a 23-mer) was designed to cut frog cyclin 1. It worked well on purified RNA and in cell-free extracts, but results in intact oocytes were variable and depended on both the amount injected and the precise protocol used. In the absence of progesterone, for example, a certain fraction of the mRNA was resistant to cutting. We interpret this to mean that not all the mRNA is accessible to the oligonucleotide and RNase H until after hormonal stimulation.

The degree of cutting of frog cyclin 2 by the antisense 23-mer proved to be far less complete than the degree of cutting of frog cyclin 1. Notwithstanding this failure to remove all cyclin mRNA, in some experiments, microinjection of the 23-mer prevented progesterone-induced maturation. Cleavage of fertilized eggs and inhibition of the periodic "surface contraction waves"

(Hara et al. 1980) were blocked more reliably. The "sense" version of the 23-mer did not block maturation or cleavage.

These tests are being repeated with oligonucleotides designed to cut both cyclin 1 and cyclin 2. It was disturbing to find experiments in which cyclin 1 was cut to completion, and yet, progesterone-induced maturation was completely unimpaired. We do not know whether this means that cyclin 1 is unimportant for maturation, whether its amino-terminal fragment (whose synthesis presumably continues) is active, or whether cyclin 2 provides a completely sufficient backup.

It is important to recognize that we have not shown that cyclin mRNA is the active principle in total frog egg poly(A)$^+$ RNA. It is quite possible that there may be other mRNAs whose translation products activate MPF, which are completely unrelated to cyclin. Pure synthetic frog cyclin mRNAs have yet to be tested for their ability to promote maturation. We note that microinjection of the mRNA for pre-MPF might not be expected to promote maturation. The oocytes already contain a pool of pre-MPF, and it is doubtful whether expansion of that pool by newly made pre-MPF could cause its activation.

Cell-free Systems from *Xenopus* Eggs Need Cyclin to Enter M Phase

Several groups have used cell-free systems from *Xenopus* eggs capable of performing multiple rounds of DNA synthesis, which also show aspects of entry into mitosis such as breakdown of the nuclear envelope and chromosome condensation (Lohka and Masui 1983; Blow and Laskey 1985; Lohka and Maller 1985; Hutchison et al. 1987; Newport and Spann 1987). Protein synthesis is required for the first entry into mitosis and for the second and subsequent rounds of DNA synthesis. To test whether cyclin synthesis accounted for the protein synthesis requirement, we added a frog cyclin-specific antisense oligonucleotide to such preparations and found that the second round of DNA synthesis and the first entry into mitosis-like state were prevented provided cyclin 1 was cut completely. No general inhibition of protein synthesis occurred under these conditions.

Cyclin Is Posttranslationally Modified and Probably Phosphorylated at the Onset of M Phase

In some sea urchin embryos, cyclin is converted into a form that migrates more slowly on one-dimensional polyacrylamide gels at around the time the fertilized eggs begin to enter mitotic prophase. A similar mobility shift was seen when sea urchin cyclin was used to promote maturation in *Xenopus* oocytes (Pines and Hunt 1987). The important point is that the new form represents a conversion of previously synthesized polypeptide and the modified form appears quite suddenly. We believe that the conversion is due to phosphorylation, because a similar shift can be seen in vitro, where newly synthesized cyclin made by translation in a reticulocyte lysate can be converted to the modified form by the addition of sea urchin extracts, ATP, and phosphatase inhibitors. The converted form can be returned to its primary form by treatment with phosphatases (L. Stewart, unpubl.). However, it has not been shown yet that the modification seen occurring in intact eggs is the same as that occurring in vitro.

When DNA synthesis is prevented by aphidicolin in fertilized sea urchin eggs, the conversion to the more slowly migrating form does not occur (T. Hunt, unpubl.). It is not clear whether the conversion is necessary to trigger mitosis or whether conversion is a consequence of entering M phase. These observations raise the possibility that cyclin may be activated by posttranslational modification and that it may not be active as a mitotic inducer in its "ground state." The identity of the activity that performs the conversion is an important question.

The Relationship between Cyclin and MPF

Entry into meiotic and mitotic prophase is accompanied by the appearance of a cytoplasmic activity known as MPF (Masui and Markert 1971; for review, see Ford 1985; Maller 1985). MPF activity appears periodically even if the egg cannot cleave, provided protein synthesis continues (Newport and Kirshner 1984; Gerhart et al. 1985). MPF activity falls precipitously at the end of metaphase (Gerhart et al. 1984). These properties are reminiscent of cyclin, and we believe that there is a close connection between cyclin and MPF, but it is not clear whether cyclin is a component of MPF, an activator of MPF, or even more remotely connected. It is still not possible to draw conclusions from the known facts. At least in oocytes, MPF exists as a precursor that can be activated without any need for protein synthesis by injection of egg cytoplasm or purified (protein) components thereof (see Maller et al., this volume). In normal oocyte maturation, however, progesterone activates the first small amount of pre-MPF by a process that absolutely requires protein synthesis (Wasserman and Masui 1975). We interpret this to mean that a translation product of a maternal mRNA is required to initiate the activation of pre-MPF. The mRNA for this hypothetical activator is presumably stored in untranslatable form in oocytes, and progesterone removes the block to its translation. However, is this hypothetical mRNA cyclin mRNA? Although we originally thought that this was almost certainly the case, because clam and urchin cyclin cause oocyte maturation and frog oocytes contain cyclin mRNA, we are puzzled by the observation that oocytes can mature with normal kinetics with essentially no frog cyclin 1 intact. Is the MPF-like activity of heterologous cyclin mRNAs perhaps misleading, or does frog cyclin 2 provide a backup? We expect that it will not be long before the sequence of purified MPF and tests with antibodies will resolve these issues. It is important to discover whether there is any cyclin *protein* in frog oocytes. J. Westendorf and J. Ruderman (unpubl.) have shown that clam oocytes contain clam cyclin B (but unlike frog oocytes, clam oocytes do not require protein synthesis to enter M phase after fertilization).

We suppose that cyclin is necessary for the appearance of MPF activity and that MPF disappears due to the destruction of cyclin. According to this model, the metaphase-anaphase transition is triggered by proteolysis. What triggers the proteolysis, and what kind of an entity performs it? How does cyclin destruction lead to MPF inactivation? The relationship between MPF and cyclin is highly pertinent to understanding the control of the cell cycle, but as usual, making connections in reverse between molecules and activities is difficult. At the moment, there does not seem to be any evidence to rule out the possibility that cyclin is an actual component of MPF. The most highly purified preparations prepared by Maller and his colleagues (this volume) contain a polypeptide in the 40-kD range, the appropriate range for cyclin, and this polypeptide appears to be a substrate for a protein kinase that migrates at about 30 kD.

Quite apart from these kinds of questions, we have no idea of the underlying enzymology of cyclin; it is surprising that none of the sea urchin or frog cyclins appear to contain protein kinase consensus sequences. One must look for other ways in which cyclin may act; e.g., perhaps it modulates the activity of a kinase or phosphatase. We expect to see light shed on these long-standing mysteries in the very near future.

ACKNOWLEDGMENTS
Work in Cambridge was supported by the Medical Research Council, the Science and Engineering Research Council and the Cancer Research Campaign; work at the Marine Biological Laboratory (Woods Hole, Massachusetts) was supported by a grant from the National Science Foundation (PCM-82-169-17) to J.V.R. The Nuffield Foundation made a grant to enable J.M. and J.P. to visit John Gerhart's laboratory in Berkeley, California, where some of these experiments were performed; they were also supported by U.S. Public Health Service grant GM-19363. We thank John Gerhart, Fred Wilt, and Andrew Murray for their generous hospitality, encouragement, and advice. Doug Melton and Colin Dingwall generously provided frog maternal mRNA cDNA libraries.

REFERENCES

Birchmeier, C., D. Broek, and M. Wigler. 1985. *RAS* proteins can induce meiosis in *Xenopus* oocytes. *Cell* **43:** 615.

Blow, J.J. and R.A. Laskey. 1985. Initiation of DNA replication in nuclei and purified DNA by a cell-free extract of *Xenopus* eggs. *Cell* **47:** 577.

Dash, P., I. Lotan, M. Knapp, E.R. Kandel, and P. Goelet. 1987. Selective elimination of mRNA *in vivo*: Complementary oligodeoxynucleotides promote RNA degradation by an RNAse H-like activity. *Proc. Natl. Acad. Sci.* **84:** 7896.

Dworkin, M.B., A. Shrutkowski, and E. Dworkin-Rastl. 1985. Mobilization of specific maternal mRNA species into polysome after fertilization in *Xenopus laevis*. *Proc. Natl. Acad. Sci.* **82:** 7636.

Evans, T., E, Rosenthal, J. Youngblom, D. Distel, and T. Hunt. 1983. Cyclin: A protein specified by maternal mRNA that is destroyed at each cleavage division. *Cell* **33:** 389.

Ford, C.C. 1985. Maturation promoting factor and cell cycle regulation. *J. Embryol. Exp. Morphol.* (Supp.) **89:** 271.

Gerhart, J.C., M. Wu, and M. Kirschner. 1984. Cell cycle dynamics of an M-phase-specific cytoplasmic factor in *Xenopus laevis* oocytes and eggs. *J. Cell Biol.* **98:** 1247.

Gerhart, J., M. Wu, M. Cyert, and M. Kirschner. 1985. M-phase promoting factors from eggs of *Xenopus laevis*. *Cytobios* **43:** 335.

Hara, K., P. Tydeman, and M. Kirschner. 1980. A cytoplasmic clock with the same period as the division cycle in *Xenopus* eggs. *Proc. Natl. Acad. Sci.* **77:** 462.

Hultin, T. 1961. The effect of puromycin on protein metabolism and cell division in fertilized sea urchin eggs. *Experientia* **17:** 410.

Hutchison, C.J., R. Cox, R.S. Drepaul, M. Gomperts, and C.C. Ford.

1987. Periodic DNA synthesis in cell-free extracts of *Xenopus* eggs. *EMBO J.* **6:** 2003.

Kuriyama, R., G.G. Borisy, and Y. Masui. 1986. Microtubule cycles in oocytes of the surf clam, *Spisula solidissimia:* An immunofluorescence study. *Dev. Biol.* **114:** 151.

Lohka, M.J. and J.L. Maller. 1985. Induction of nuclear envelope breakdown, chromosome condensation, and spindle formation in cell-free extracts. *J. Cell. Biol.* **101:** 518.

Lohka, M.J. and Y. Masui. 1983. Formation in vitro of sperm pronuclei and mitotic chromosomes induced by amphibian ooplasmic contents. *Science* **220:** 719.

Maller, J.L. 1985. Egg maturation in amphibians. In *Developmental biology: A comprehensive synthesis. Oogenesis.* (ed. L.W. Broder), vol. 1, p. 289. Plenum Press, New York.

Maller, J.L. and E.G. Krebs. 1980. Regulation of oocyte maturation. *Curr. Top. Cell. Regul.* **16:** 271.

Masui, Y. and C.L. Markert. 1971. Cytoplasmic control of nuclear behavior during meiotic maturation of frog oocytes. *J. Exp. Zool.* **177:** 129.

Minshull, J. and T. Hunt. 1986. The use of single-stranded DNA and RNase H to promote quantitative "hybrid arrest of translation" of mRNA/DNA hybrids in reticulocyte cell-free translations. *Nucleic Acids Res.* **14:** 6433.

Newport, J.W. and M. Kirschner. 1984. Regulation of the cell cycle during early *Xenopus* development. *Cell* **37:** 731.

Newport, J.W. and T. Spann. 1987. Disassembly of the nucleus in mitotic extracts: Membrane vesicularization, lamin disassembly and chromosome condensation are independent processes. *Cell* **48:** 219.

Picard, A., G. Peaucellier, F. Le Bouffant, C. Le Peuch, and M. Dorée. 1985. Role of protein synthesis and proteases in production and inactivation of maturation-promoting activity during meiotic maturation of starfish oocytes. *Dev. Biol.* **109:** 311.

Pines, J. and T. Hunt. 1987. Molecular cloning and characterization of the mRNA for cyclin from sea urchin eggs. *EMBO J.* **6:** 2987.

Rosenthal, E.T., T.R. Tansey, and J. Ruderman. 1983. Sequence-specific adenylations and deadenylations accompany changes in the translation of maternal messenger RNA after fertilization in *Spisula* oocytes. *J. Mol. Biol.* **166:** 309.

Shuttleworth, J. and A. Colman. 1988. Antisense oligonucleotide-directed cleavage of mRNA in *Xenopus* oocytes and eggs. *EMBO J.* **7:** 427.

Standart, N., J. Minshull, J. Pines, and T. Hunt. 1987. Cyclin synthesis, modification and destruction during meiotic maturation of the starfish oocyte. *Dev. Biol.* **124:** 248.

Swenson, K.I., K.M. Farrell, and J.V. Ruderman. 1986. The clam embryo protein cyclin A induces entry into M-phase and the resumption of meiosis in *Xenopus* oocytes. *Cell* **47:** 861.

Wagenaar, E.B. 1983. The timing of synthesis of protein required for mitosis in the cell cycle of the sea urchin embryo. *Exp. Cell. Res.* **144:** 3393.

Wasserman, W.J. and Y. Masui. 1975. Effects of cycloheximide on a

cytoplasmic factor initiating meiotic maturation in *Xenopus* oocytes. *Exp. Cell. Res.* **91:** 381.
Wilt, F.H., H. Sakai, and D. Mazia. 1967. Old and new proteins in the formation of the mitotic apparatus in cleaving sea urchin eggs. *J. Mol. Biol.* **27:** 1.
Woodland, H.R. 1982. The translational phase of early development. *Biosci. Rep.* **2:** 471.

Mitotic Cyclins and Regulation of the Cell Cycle in Early Embryos

J. Ruderman

Department of Zoology, Duke University
Durham, North Carolina 27706

Oocytes contain enormous stockpiles of mitotic proteins that enable early embryos to proceed through a series of very rapid cell-division cycles. Despite this, the oocyte is arrested at a particular stage of the cell cycle, with the exact point and the external release signal varying among different species. In clams, for example, interaction with the sperm sets off a rapid series of ionic changes and triggers the synthesis of a small number of additional proteins. These newly made embryonic proteins are encoded by maternal mRNAs that are stored in the oocyte and begin to be translated within minutes of fertilization. Some of these newly synthesized proteins enable the zygote to proceed beyond S phase and into the first mitosis. Furthermore, there is a recurrent need for protein synthesis in each cell cycle, indicating that newly made mitosis-inducing proteins are destroyed or inactivated at the end of each mitosis and then resynthesized in the next cell cycle (Wilt et al. 1967; Rosenthal et al. 1980, 1983; Evans et al. 1983; Wagenaar 1983; Swenson et al. 1986; Pines and Hunt 1987).

Among the most abundant maternal mRNAs activated by fertilization of clam and sea urchin oocytes are those encoding the cyclins, proteins whose levels rise and fall during each cell cycle and appear to regulate entry into mitosis and exit from mitosis into the next interphase (Evans et al. 1983; Swenson et al. 1986; Westendorf and Ruderman 1986; Pines and Hunt 1987). In clam embryos, for example, cyclin A protein is absent in the oocyte and first appears about 10 minutes after fertilization; there is a small pool of cyclin B protein in oocytes, and additional cyclin B protein is made soon after fertilization (Fig. 1). Thereafter, both cyclin proteins are made continuously and accumulate in S phase and early M phase; they are rapidly destroyed near the end of each mitosis. Cyclin A disappears slightly ahead of the metaphase-anaphase transition; this is

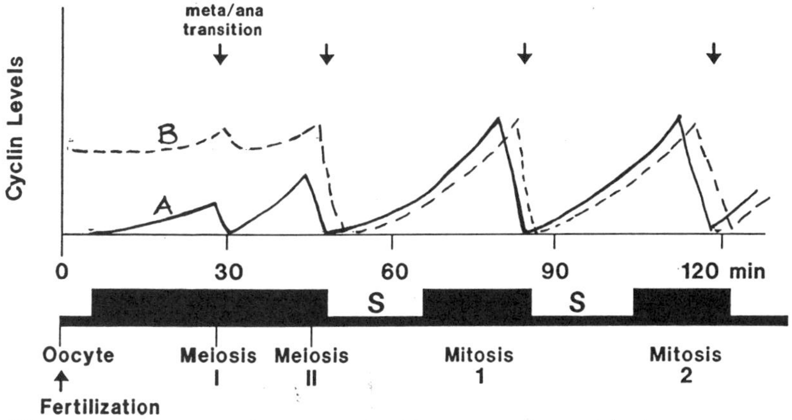

Figure 1 The periodic rise and fall in cyclin levels is due to continuous synthesis across the cell cycle, followed by periodic destruction near the end of each M phase.

followed by the disappearance of cyclin B a few minutes later, just at the metaphase-anaphase transition.

Earlier work had shown that the rise of cyclin A levels leads to entry into M phase of the cell cycle, suggesting that cyclin A acts as a mitotic inducer. Pure clam cyclin A mRNA (prepared using an RNA transcription vector) was injected into *Xenopus* oocytes, which are arrested naturally at the G_2/M border of meiosis I. This mRNA was translated, and the resulting cyclin A protein led to entry into meiotic M phase, as indicated by breakdown of the nuclear envelope, chromosome condensation, formation of the meiotic spindle, and migration of the spindle to the animal pole cortex (Swenson et al. 1986). The major sea urchin cyclin protein promotes the same events in frog oocytes (Pines and Hunt 1987). These results indicate that cyclin A, which cycles during both the meiotic and mitotic divisions, has a direct and natural role in promoting entry into M phase.

In contrast, several indirect experiments suggest that high levels of cyclin B protein appear to keep cells in M phase; it is probably the destruction of cyclin B, that allows cells to exit from mitosis into interphase of the next cell cycle. Injection of clam cyclin B mRNA into frog oocytes is not followed by the resumption of meiosis (Westendorf and Ruderman 1987 and 1986). When clam embryos are exposed to colchicine, they arrest in M phase for several hours; cyclin A falls on schedule, but cyclin B stays high. If the protein synthesis inhibitor emetine is subsequently given to these metaphase-arrested embryos, cyclin B disappears about 1 hour later. The disappearance of cyclin B shows a close temporal correlation with abortive cleavage attempts, which are rapidly followed by exit into the next interphase (T. Hunt and J. Ruderman, in prep.). We are now trying to test the mitotic maintenance role of cyclin B directly by asking whether high levels of cyclin B introduced into cleaving frog blastomeres can arrest those cells in M phase.

Comparisons of the predicted amino acid sequences of clam cyclin A and B (Swenson et al. 1986; J. Westendorf and J. Ruderman, in prep.), sea urchin, and frog cyclin (Pines and Hunt 1987; J. Minshull and T. Hunt, in prep.) show that all are about 400–430 amino acids long and are very similar in the middle third, a region loosely termed the "cyclin box." The amino acid sequence of cyclin B has been the most informative one so far. Unlike the other three, cyclin B contains kinase consensus sequences typical of both the ATP-binding site and a potential phosphoacceptor region; it also contains stretches of amino acids, suggesting binding sites for calcium and zinc (Westendorf and Ruderman 1988 and in prep.). Immunoprecipitates of cyclin B from M-phase embryos have considerable kinase activity in vitro; the phosphorylated reaction products include cyclin B itself and a small number of other proteins. The substrate specificity of cyclin B kinase is affected by the addition of zinc, which favors the autophosphorylation event. In the in vitro assay, cyclin B phosphorylates exogenous histone H1 and myosin light chain but has no activity toward casein. This activity of cyclin B is particularly interesting because of the close relationship between the phosphorylation of many mitotic proteins (including MPF) and entry to mitosis and by the finding that the in vitro kinase activity of the *cdc28* gene protein is stimulated by zinc (Reed et al. 1985; Mendenhall et al. 1987).

Curiously, immunoprecipitates of cyclin A—which shows no recognizable kinase amino acid consensus sequences itself—have considerable kinase activity in vitro. In the absence of zinc, cyclin A is the only detectable phosphorylated product. In the presence of zinc, we see the appearance of a phosphoprotein that migrates about 5 kD above the 55-kD cyclin protein (K. Swenson and J. Ruderman, in prep.). At this point, we cannot say whether cyclin A itself is a kinase or a substrate of a kinase that coprecipitates with cyclin A. Unfortunately, attempts to study the kinase activities using cloned cyclin A and B mRNAs translated in reticulocyte lysates have not been reliably successful.

Recently, we have been able to develop an in vitro system (made from lysates of early M-phase clam embryos) in which both cyclin A and B show the scheduled and slightly offset destruction that is characteristic of their disappearance at the end of mitosis in vivo. This in-vitro-scheduled cyclin destruction is very reproducible, and the activity survives multiple rounds of freezing and thawing. Very preliminary experiments show that the in vitro destruction of the cyclins is not affected by the addition of calcium, EGTA, or several protease inhibitors, including aprotinin and α2-macroglobulin, but is blocked by EDTA, zinc, and antipain (a thiol-specific protease inhibitor). Some of the lysates show the further property that destruction of the cyclins is followed by their reappearance. The addition of the protein synthesis inhibitor emetine has no effect on the disappearance of the cyclins but does prevent their reappearance (Luca and Ruderman 1988). We are now trying to use this in vitro system with SP6 RNAs made from cloned deletion and insertion mutations of the cyclins to map the sites that are responsible for their periodic stabilization and destruction. These sorts of altered cyclins will also be useful for identifying the regions of cyclin A responsible for its ability to induce M phase and the regions necessary for the kinase activities of the cyclins.

What is the relationship between the cyclins and maturation-promoting factor (MPF), a ubiquitous M-phase-promoting activity that rises and falls during the cell cycle, with kinetics that are remarkably like those of the cyclins? Recent work (Cyert and Kirschner 1988; Dunphy and Newport 1988) has revealed that MPF is reversibly inactivated, not destroyed, as cells leave M phase and go into the next interphase. In contrast, all available evidence argues that both cyclins A and B are proteolytically destroyed at the end of each mitosis. At this point, then, most evidence supports the idea that the cyclins act as regulators of MPF activity. This is a prediction that is now ready for an experimental test.

REFERENCES

Cyert, M. and Kirschner. 1988. Regulation of MPF activity *in vitro*. *Cell* **53:** 185.

Dunphy, W. and J. Newport. 1988. Mitosis-inducing factors are pres-

ent in a latent form during interphase in the *Xenopus* embryo. *J. Cell Biol.* (in press).
Evans, T., E. Rosenthal, T. Youngblom, D. Distel, and T. Hunt. 1983. Cyclin: A protein specified by maternal mRNA that is destroyed at each cleavage division. *Cell* **33:** 389.
Luca, F. and J.V. Ruderman. 1988. Periodic stabilization and destruction *in vitro* of mitotic cyslins from early embryos. *ICN-UCLA Symp. Mol. Cell Biol.* (in press).
Mendenhall, M.D., C.A. Jones, and S.I. Reed. 1987. Dual regulation of the yeast CDC28-p40 protein kinase complex: Cell cycle, pheromone and nutrient limitation effects. *Cell* **50:** 927.
Pines, J. and T. Hunt. 1987. Molecular cloning and characterization of the mRNA for cyclin from sea urchin eggs. *EMBO J.* **6:** 2987.
Reed, S.I., J.A. Hadwiger, and A.T. Lorincz. 1985. Protein kinase activity associated with the product of the yeast cell division cycle gene *CDC28*. *Proc. Natl. Acad. Sci.* **82:** 4055.
Rosenthal, E.T., T. Hunt, and J.V. Ruderman. 1980. Selective translation of mRNA controls the pattern of protein synthesis during early development of the clam, *Spisula*. *Cell* **20:** 487.
Rosenthal, E.T., T.R. Tansey, and J.V. Ruderman. 1983. Sequence-specific adenylations and deadenylations accompany changes in the translation of maternal mRNA after fertilization of *Spisula* oocytes. *J. Mol. Biol.* **166:** 309.
Swenson, K.I., K.M. Farrell, and J.V. Ruderman. 1986. The clam embryo protein cyclin A induces entry into M phase and the resumption of meiosis in *Xenopus* oocytes. *Cell* **47:** 861.
Wagenaar, E.B. 1983. The timing of synthesis of proteins required for mitosis in the cell cycle of the sea urchin embryo. *Exp. Cell Res.* **144:** 393.
Westendorf, J.M. and J.V. Ruderman. 1986. Molecular cloning and antibody studies of cyclin B, a protein involved in the cell cycle. *J. Cell. Biol.* **103:** 555a.
———. 1988. Clam cyclin B is a kinase. *J. Cell Biol.* **105:** 567a.
Wilt, F.H., H. Sakai, and D. Mazia. 1967. Old and new protein in the formation of the mitotic apparatus in cleaving sea urchin eggs. *J. Mol. Biol.* **27:** 1.

Two Genes, *nimA* and *bimE*, That Regulate Mitosis in *Aspergillus nidulans*

N.R. Morris, S.A. Osmani, D.B. Engle, and R.T. Pu

Department of Pharmacology
University of Medicine and Dentistry of New Jersey
Robert Wood Johnson Medical School, Piscataway, New Jersey 08854

During the 1970s, several laboratories, including ours, initiated a new approach to the study of mitosis by isolating cell-cycle mutants of microbial eukaryotes, most usefully in *Saccharomyces cerevisiae* (Hartwell et al. 1973), *Schizosaccharomyces pombe* (Nurse et al. 1976), and *Aspergillus nidulans* (Morris 1976). With the advent of DNA-mediated transformation of these organisms, many of the cell-cycle genes were cloned by complementation of their mutant phenotypes, and their gene products were characterized. We now know, particularly from the work of P. Nurse and his colleagues, that protein kinases are important in the regulation of mitosis (Simanis and Nurse 1986; Russell and Nurse 1987a,b). The amino acid sequence of one of these protein kinases, the *CDC2* protein of *S. pombe* (Hindley and Phear 1984), is highly conserved during evolution and has been identified in the human genome (Draetta et al. 1987; Lee and Nurse 1987). In this paper, we describe two genes, *nimA* and *bimE*, that are involved in mitotic regulation in *A. nidulans*. The *nimA* gene product is a positive inducer of chromatin condensation and spindle polymerization, which appears to be a previously undescribed protein kinase. The *bimE* gene product appears to prevent mitosis from occurring during interphase.

RESULTS

The *nimA* Gene

The *nimA5* mutation of *nimA* causes a temperature-sensitive cell-cycle block at the G_2–M boundary. Cycling nuclei accumulate at this boundary but then enter mitosis rapidly and synchronously when downshifted to permissive temperature (Oakley and Morris 1983; Osmani et al. 1987). Because the *nimA5* mutation is recessive, it most likely represents a loss of gene function. We therefore conclude that the *nimA* gene product is required for entry into mitosis.

The wild-type *nimA* gene was cloned from a plasmid genomic library by DNA-mediated transformation and complementation of the temperature sensitivity of *nimA5* (Osmani et al. 1987). On Northern analysis, it hybridizes to a 3.6-kb mRNA, whose amount fluctuates according to the cell cycle. The *nimA* message is low during S, rises during G_2, reaches a peak at M, and drops precipitously after M phase. When cells are blocked in interphase at S with the DNA synthesis inhibitor hydroxyurea, nuclei do not enter mitosis and the *nimA* mRNA level remains low. When cells are mitotically blocked with benomyl, *nimA* mRNA rises to a high level as nuclei enter mitosis and remains high as long as mitosis is prolonged. Thus, the level of *nimA* mRNA (and presumably the *nimA* gene product) correlates with the mitotic state of the nucleus.

The *nimA* genomic clone was used to identify *nimA* cDNA clones from a λgt10 cDNA library. A full-length clone was sequenced and found to have the amino acid signature of a protein kinase (Osmani et al. 1988a). Except for this protein kinase consensus sequence, *nimA* does not resemble any of the protein kinase genes of *S. pombe* or *S. cerevisiae* published thus far (nor does it resemble cyclin). The *CDC2* gene of *S. pombe* and the *CDC28* gene of *S. cerevisiae* code for functions similar to that of *nimA*, except that unlike *nimA*, these functions are also required at the G_1/S transition. A comparison of the amino acid sequences of *nimA*, *CDC2*, *CDC28*, and a human homolog of *CDC28* reveals that *CDC2*, *CDC28*, and the human gene are amazingly similar, whereas *nimA* is clearly divergent.

Not only does the level of *nimA* correlate with mitosis, as noted above, but deliberately raising the level of *nimA* causes cells to enter mitosis. The *nimA* genomic clone was fused to an upstream, inducible *alcA* (alcohol dehydrogenase) promoter and reintroduced into *Aspergillus* by transformation. Shifting transformed strains from noninducing to inducing medium causes overproduction of *nimA* mRNA and is lethal in strains containing more than one or two extra copies of the *alcA-nimA* construct. Induction halts nuclear division, causes a rapid and persistent increase in mitotic index, causes a persistent loss of the cytoplasmic interphase microtubule pattern, and, in some cases, causes nuclei to appear "pulverized." Overproduction of *nimA* induces mitosis even in cells that are in interphase.

When cells are arrested in S phase with hydroxyurea, overproduction of *nimA* induces nuclei to enter and be blocked in mitosis. Thus, the elevation of *nimA* appears to be the only necessary requirement for induction of mitosis in *A. nidulans* (Osmani et al. 1988a).

The *bimE* Gene

The temperature-sensitive *bimE7* mutation causes cells to become blocked in mitosis at restrictive temperature, with condensed nuclear chromatin and a medial mitotic spindle. The wild-type *bimE* gene has been cloned from a genomic plasmid library by DNA-mediated transformation and complementation of the temperature-sensitive *bimE7* phenotype. Northern analysis showed that this gene recognizes a single 7.0-kb mRNA. A 7-kb cDNA clone has been isolated from a λgt10 cDNA library and is in the process of being sequenced (D. Engle et al., in prep.).

We initially believed that the *bimE7* mutation caused a lesion in some protein required for the normal functioning of the mitotic spindle (by analogy with the effect of spindle poisons, which cause a prolongation of chromatin condensation while spindle function is blocked). However, it now seems likely that *bimE* is a component of a regulatory system that prevents mitosis from occurring during interphase, possibly by down-regulating *nimA*. The first clue that *bimE* might be regulatory came from construction of a *bimE7 nimA5* double mutant (Osmani et al. 1988b). Because *nimA5* causes a G_2 block and *bimE7* causes an M block, we expected that the double mutant would be blocked in G_2 at restrictive temperature. Instead, it became blocked in M but was delayed by 1 hour in its entry into mitosis (in comparison with a blocked strain carrying only *bimE7*). The *nimA* mRNA also increased, in this case, with the same kinetics as in a control strain carrying only the *bimE7* mutation (S.A. Osmani and N.R. Morris, unpubl.). We interpreted this unexpected result to mean that inactivation of the recessive *bimE7* gene caused an increase in *nimA5* mRNA (and presumably *nimA5* gene product), sending the nuclei into M. We supposed that the delay in entering mitosis was caused by *nimA5* being leaky, with only a fraction of the protein kinase activity of *nimA*$^+$. To determine whether inactivation of *bimE* caused mitosis to occur only when cells were already in G_2 or whether mitosis could be induced earlier during interphase, we blocked a strain carrying *bimE7* in S with hydroxyurea and then shifted it to restrictive temperature. The result was that the mitotic index increased and the nuclei became blocked in M, despite the fact that cells blocked in S phase normally do not enter mitosis. A null mutation of the *bimE* gene product generated by an integrative gene disruption also caused cells to become blocked in M. Thus, *bimE* appears to be a negative regulator that keeps cells from entering mitosis during interphase (Osmani et al. 1988b).

Significance

Not only are there a number of protein kinases involved in regulating mitosis in simple eukaryotes, but there is also an extensive biochemical literature linking protein phosphorylation and mitosis in higher eukaryotes; in addition, a number of potential protein targets for kinase activity have been delineated (see, e.g., Laskey 1983; Centonze et al. 1986). Because there is a human gene that complements and is similar in sequence to the *cdc2* mutation of *S. pombe*, we believe that other genes involved in mitosis, e.g., *nimA*, may also be found in higher eukaryotes. We are intrigued by the functional resemblance between the *nimA* putative protein kinase and the maturation-promoting factor (MPF), a protein that induces mitosis in vertebrate cells. Highly purified (albeit not pure) MPF has protein kinase activity (Lohka et al. 1987), and its activity is cyclical and tightly coupled to the cell cycle, like *nimA* rising in G_2 and peaking at M (Gerhart et al. 1984).

We wish to emphasize that *nimA* and *bimE* are not the respective *Aspergillus* homologs of *S. pombe CDC2* and *WEE1*, as shown by comparing the *nimA* sequence with that of *CDC2* and by differences between mutant phenotypes. The temperature-sensitive mutation *nimA5* arrests only at the G_2/M boundary, whereas *cdc2* also arrests at the G_1/S boundary. Overproduction of wild-type *nimA* causes persistent chromatin condensation and mitotic spindle formation, but overproduction of *CDC2* has no effect. Inactivation of *bimE* causes the same phenotype as overproduction of *nimA*: Inactivation of *WEE1* causes a "wee" phenotype. It should also be noted that to our knowledge, among the fungal genes thought to regulate mitosis, *nimA* is the only one that appears to be under cell-cycle control.

Clearly, future directions to our work will be to identify the *nimA* protein, to see whether it has kinase activity, to identify its protein targets, and to determine whether phosphorylation of these targets causes mitotic events. We will also pursue the factors responsible for the cyclical expression of *nimA*, particularly *bimE* and other genes with which it may interact.

ACKNOWLEDGMENT

This work was supported by National Institutes of Health grants GM-29228 and GM-34711.

REFERENCES

Centonze, V.E., D.D. Vandre, and G.G. Borisy. 1986. Growth of microtubules on mitotic centrosomes is blocked by MPM-2. *J. Cell Biol.* **103:** 412 (Abstr.).

Draetta, G., L. Brizuela, J. Potashkin, and D. Beach. 1987. Identification of p34 and p13, human homologs of the cell cycle regulators of fission yeast encoded by *cdc2*$^+$ and *suc1*$^+$. *Cell* **50:** 319.

Gerhart, J., M. Wu, and M. Kirschner. 1984. Cell cycle dynamics of an M-phase-specific cytoplasmic factor in *Xenopus laevis* oocytes and eggs. *J. Cell Biol.* **98:** 1247.

Hartwell, L.H., R.K. Mortimer, J. Culotti, and M. Culotti. 1973. Genetic control of the cell division cycle in yeast: V. Genetic analysis of *cdc* mutants. *Genetics* **74:** 267.

Hindley, J. and G.A. Phear. 1984. Sequence of the cell division gene *CDC2* from *Schizosaccharomycetes pombe:* Patterns of splicing and homology to protein kinases. *Gene* **31:** 129.

Laskey, R.A. 1983. Phosphorylation of nuclear proteins. *Philos. Trans. R. Soc. Lond. B.* **302:** 143.

Lee, M.G. and P. Nurse. 1987. Complementation used to clone a human homologue of the fission yeast cell cycle control gene *cdc2*. *Nature* **327:** 31.

Lohka, M.J., M. Hayes, and J.L. Maller. 1987. A protein kinase associated with highly purified maturation promoting factor. *J. Cell Biol.* **105:** 103 (Abstr.).

Morris, N.R. 1976. Mitotic mutants of *Aspergillus nidulans*. *Genet. Res.* **26:** 237.

Nurse, P., P. Thuriaux, and K. Nasmyth. 1976. Genetic control of the cell division cycle in the fission yeast *Schizosaccharomyces pombe*. *Mol. Gen. Genet.* **146:** 167.

Oakley, B.R. and N.R. Morris. 1983. A mutation in *Aspergillus nidulans* that blocks the transition from interphase to prophase. *J. Cell Biol.* **96:** 1155.

Osmani, S.A., G.S. May, and N.R. Morris. 1987. Regulation of the mRNA levels of *nimA*, a gene required for the G_2-M transition in *Aspgergillus nidulans*. *J. Cell Biol.* **104:** 1495.

Osmani, S.A., R.T. Pu, and N.R. Morris. 1988a. Mitotic induction and maintenance by overexpression of a G_2 specific gene that encodes a potential protein kinase. *Cell* **53:** 237.

Osmani, S.A., D. Engle, J.H. Doonan, and N.R. Morris. 1988b. Spindle formation and chromatin condensation in cells blocked at interphase by mutation of a negative cell cycle control gene. *Cell* **52:** 241.

Russell, P. and P. Nurse. 1987a. Negative regulation of mitosis by *wee1*, a gene encoding a protein kinase homolog. *Cell* **49:** 559.

———. 1987b. The mitotic inducer *nim1* functions in a regulatory network of protein kinase homologs controlling the initiation of mitosis. *Cell* **49:** 569.

Simanis, V. and P. Nurse. 1986. The cell cycle control gene *cdc2* of fission yeast encodes a protein kinase potentially regulated by phosphorylation. *Cell* **45:** 261.

RAD9 Gene Function Controls the Cell-cycle Response to DNA Damage in the Yeast *Saccharomyces cerevisiae*

T.A. Weinert and L.H. Hartwell

Department of Genetics, SK-50, University of Washington
Seattle, Washington, 98195

Cell division is controlled in response to a variety of physiological changes, including damage to DNA. Bacteria and animal cells respond to DNA damage by delaying cell division. The purpose of division delay in bacteria has not been demonstrated clearly, although the genetic and molecular basis of the response is well understood; DNA damage activates the *recA* protease, leading to synthesis of an inhibitor of cell septation (Little and Mount 1982). The nature and purpose of DNA-damage-sensitive division delay in eukaryotes are understood, although underlying mechanisms remain obscure. DNA damage induces arrest in the G_2 phase of the cell cycle, after DNA replication and before mitosis. A number of observations (see, e.g., Lau and Pardee 1982) suggest that G_2 arrest may simply provide the cell ample time to repair DNA lesions and thus ensure integrity of condensed chromosomes for segregation at mitosis.

Common features of division delay in many diverse cell types, including sea urchin eggs and yeast (Burns 1956; Rustad 1971; Brunborg and Williamson 1978), suggest the existence of common molecular mechanisms for G_2 delay in eukaryotes. Yeast cells treated with X-irradiation or methylmethane sulfonate delay cell division when the nucleus is at the neck of a large budded cell (Burns 1956; Brunborg and Williamson 1978; Kupiec and Simchen 1985), a phenotype identical to that of several mutants that block nuclear division in S or G_2 at their restrictive temperatures. Burns (1956) and Brunborg and Williamson (1978) also provided evidence that yeast cells irradiated in G_2 delay in G_2 before resuming cell division. We confirm and extend observations on X-irradiated yeast cells by demonstrating that wild-type haploid yeast cells irradiated with X-rays either arrest or delay cell division in the G_2 phase of the cell cycle. Irradiated G_1- and M-phase haploid cells arrest irreversibly in G_2 and die, whereas irradiated G_2-phase haploid cells delay in G_2 for a time proportional to the extent of damage before resuming cell division.

We have investigated the genetic basis for G_2 delay in the yeast *Saccharomyces cerevisiae* by investigating division delay in existing radiation-sensitive (*rad*) mutants. Most *rad* mutants (*rad6, rad8, rad10, rad11, rad12, rad18,* and *rad52*) display normal cell-cycle arrest in response to X-irradiation. In contrast, *rad9* cells irradiated in any phase of the cycle do not delay cell division in G_2 but continue to divide for several generations and die. Analysis of nuclear morphology corroborates that *rad9* cells are defective in division arrest. Three hours postirradiation, *RAD*+ and *rad52* cells showed high percentages of cells with the medial nuclear division morphology expected of cells arrested in G_2, whereas *rad9* cells treated similarly showed less of an increase in cells with this morphology.

The *RAD9* gene product might have functions in addition to cell-cycle delay, e.g., directly as a repair enzyme or indirectly as a controlling element of repair enzyme expression or activity. If the *RAD9* gene product only controls cell-cycle progression, the radiation sensitivity of *rad9* cells might be suppressible by drug-induced G_2 delay. To test this idea, cells were arrested with methyl benzimidazole carbamate (MBC) and irradiated, and the MBC block was maintained to simulate a G_2 delay response. *rad9* cells arrested in MBC before irradiation showed a dramatic increase in cell viability but only when left in MBC for 4 hours postirradiation. These results demonstrate that (1) the *RAD9* function is not required for DNA repair, provided cells are blocked in G_2, and (2) delay in G_2 is necessary for the viability of cells with DNA damage, as observed in mammalian cells. Similar results have been obtained with UV-irradiated cells, demonstrating that the UV sensitivity of *rad9* mutant cells is probably also due to a failure of G_2 delay. We conclude that the *RAD9*-dependent response detects potentially lethal DNA damage and arrests the cell cycle in G_2 until such damage is repaired.

We were interested in determining whether *RAD9* played any role in the normal cell cycle or was employed only in response to extrinsically induced DNA damage. The *RAD9* gene was cloned, and a deletion of the chromosomal copy was created. Cells with the deletion allele grew at normal rates, demonstrating that the gene product is not essential for mitosis. However, *rad9* mutant cells disomic for chromosome 7 were found to lose one copy of chromosome 7 at a rate tenfold greater than that of *RAD9* cells; thus, the *RAD9* function is important for the repair of some spontaneous chromosome damage.

A more dramatic demonstration of the role of *RAD9* in intrinsically generated damage came from combining *rad9* with certain *cdc* mutations that cause DNA damage and irreversible cell-cycle arrest in G_2. *rad9* mutation is lethal with *cdc9* (DNA ligase) and suppresses the division arrest and increases lethality of *cdc17* (DNA polymerase I) and *cdc13* at the restrictive temperature.

We suggest that *RAD9* imposes a checkpoint in the normal cell cycle whose function is to assure that DNA replication is complete before the cell is allowed to enter mitosis. This function may be one of a number that distinguishes the somatic from the embryonic cell cycle.

REFERENCES

Brunborg, G. and D.H. Williamson. 1978. The relevance of the nuclear division cycle to radiosensitivity in yeast. *Mol. Gen. Genet.* **162:** 277.

Burns, V.W. 1956. X-ray induced division delay of individual yeast cells. *Radiat. Res.* **4:** 394.

Kupiec, M. and G. Simchen. 1985. Arrest of the mitotic cell cycles and of meiosis in *Saccharomyces cerevisiae* by MMS. *Mol. Gen. Genet.* **201:** 558.

Lau, C.C. and A.B. Pardee. 1982. Mechanism by which caffeine potentiates lethality of nitrogen mustard. *Proc. Natl. Acad. Sci.* **79:** 2942.

Little, J.W. and D.W. Mount. 1982. The SOS regulatory system of *Escherichia coli*. *Cell* **29:** 11.

Rustad, R.C. 1971. Radiation responses during the mitotic cycle of the sea urchin egg. In *Developmental aspects of the cell cycle* (ed. I.L. Cameron et al.), p. 127. Academic Press, New York.

The Ubiquitin System and Cell-cycle Control

S. Jentsch,[1] J.P. McGrath,[2] and A. Varshavsky[2]

[1]Friederich-Miescher-Laboratorium der Max-Planck-Gesellschaft
7400 Tübingen, Federal Republic of Germany

[2]Department of Biology, Massachusetts Institute of Technology
Cambridge, Massachusetts 02139

Ubiquitin, a small, highly conserved eukaryotic protein exists in cells either free or covalently joined to a variety of cellular proteins. Biochemical and genetic evidence indicates that conjugation of ubiquitin to intracellular proteins is essential for their selective degradation (see Finley and Varskavsky 1985; Hershko and Ciechanover 1986). An alternative, non-proteolytic role of ubiquitin, in which its reversible joining to an acceptor protein could modulate protein function without metabolically destabilizing the acceptor protein, has also been suggested, particularly because of the presence of the ubiquitinated histones H2A and H2B in chromatin. All known functions of ubiquitin are mediated by its joining to cellular proteins. The enzymatic reactions involved in the formation of ubiquitin conjugates have been studied extensively with enzymes isolated from mammalian cells (Hershko et al. 1983; Pickart and Rose 1985).

In an initial ATP-requiring step, the carboxyl terminus of ubiquitin is joined, through a thiolester bond, to a cysteine residue of the ubiquitin-activating enzyme, E1. The activated ubiquitin is then transferred to specific cysteine residues in several smaller proteins, collectively referred to as ubiquitin-conjugating (E2) enzymes. These proteins then donate ubiquitin to specific protein substrates, forming branched conjugates in which the carboxyl terminus of ubiquitin is joined by an isopeptide bound to ε-amino groups of lysine residues in acceptor proteins. Recently, we have purified several components of the ubiquitin-protein ligase system from the yeast *Saccharomyces cerevisae* (Jentsch et al. 1987) and have undertaken the cloning of the corresponding genes.

The Ubiquitin-activating Enzyme and Cell-cycle Control

The temperature-sensitive mouse cell line *ts*85 arrests at the nonpermissive temperature in the late S and early G_2 phase of the cell cycle. Arrested cells are defective in chromosome condensation and histone H1 phosphorylation, and they also lack the ubiquitin-H2A semihistone, uH2A. Moreover, arrested cells display strikingly reduced degradation of the otherwise short-lived intracellular proteins. The conditional cell-cycle arrest in *ts*85 cells was shown to be conferred by their temperature-sensitive ubiquitin-activating enzyme (for review, see Finley and Varshavsky 1985). To allow a deeper genetic and biochemical analysis of the E1 enzyme, we purified the yeast homolog of this enzyme and raised antibodies against it. Using affinity-purified anti-E1 antibodies, we cloned the corresponding gene (*UBA1*, for *ub*iquitin-*a*ctivating enzyme) and showed that it encodes a protein of approximately 115 kD (J.P. McGrath et al., unpubl.). Deletion of this gene and generation of temperature-sensitive alleles will enable us to address the cell-cycle function of *UBA1* and will also help in defining other functions of the ubiquitin system.

The Yeast DNA Repair Gene *RAD6* Encodes a Ubiquitin-conjugating Enzyme

One of the isolated yeast ubiquitin-conjugating enzymes ($E2_{20K}$) catalyzes the transfer of ubiquitin to histone H2A and H2B in the presence of ATP and purified ubiquitin-activating enzyme, E1. Recently, we showed that the $E2_{20K}$ enzyme is the product of the yeast DNA repair gene, *RAD6* (Jentsch et al. 1987). When supplemented with ATP and E1 enzyme, extracts of *Escherichia coli* cells expressing the cloned yeast *RAD6* gene mediated covalent conjugation of ubiquitin to histones H2A and H2B (Jentsch et al. 1987). The enzymatic activity of the *RAD6* gene product suggests that at least some of the functions attributed to this gene are mediated by ubiquitination of specific target proteins. In *S. cerevisiae*, the *RAD6* gene (renamed *UBC2*, for *ub*iquitin-*c*onjugating enzyme) plays a central role in one of the three epistasis groups involved in DNA repair (Friedberg 1983). The *RAD6(UBC2)* gene is required for a variety of cellular functions, including DNA repair, induced mutagenesis, and sporulation. The in vitro specificity of the *RAD6(UBC2)* protein suggests that its in vivo substrates may include basic proteins such as chromosomal histones.

The Yeast Cell-cycle Gene *CDC34* Encodes a Ubiquitin-conjugating Enzyme

Mutants in the gene *CDC34* of the yeast *S. cerevisiae* are defective in the transition from the G_1 to the S phase of the cell cycle (Byers 1981). Recently, the *CDC34* gene was cloned and was shown to encode a protein whose amino acid sequence is similar to that of the *RAD6(UBC2)* protein (M.G. Goebl et al., pers. comm.). By expressing the cloned *CDC34* gene in *E. coli*, we showed that *CDC34* catalyzes the covalent attachment of ubiquitin to histones H2A and H2B in vitro (see Fig. 1A) (M.G. Goebl et al., in prep.). In addition to their similar substrate requirements in vitro, both the *RAD6(UBC2)* and *CDC34* (renamed *UBC3*) enzymes have highly acidic carboxy-terminal regions, which are absent from the other sequenced ubiquitin-conjugating enzyme, *UBC1* (S. Jentsch et al., unpubl.). By deleting the acidic region of *CDC34(UBC3)*, we demonstrated that the truncated protein (*cdc34-3)* can still form a thiolester with ubiquitin in the presence of ATP and E1 enzyme but is deficient in ubiquitin-histone conjugation (see Fig. 1) (M.G. Goebl et al., in prep.). These results suggest that at least the in vitro substrate specificity of *CDC34(UBC3)* enzyme is due, in part, to ionic interactions between its carboxyl terminus and its basic substrates.

DISCUSSION

The finding that the yeast DNA repair gene *RAD6* and the yeast cell-cycle gene *CDC34* each encode distinct ubiquitin-conjugating enzymes underscores further the remarkable functional diversity of the ubiquitin system. If histones will indeed prove to be among the relevant in vitro substrates of both en-

Figure 1 The *CDC34*-gene product is a ubiquitin-conjugating enzyme. Extracts of *E. coli* cells expressing the cloned yeast *CDC34* gene (lanes *1, 2,* and *5*) of nonexpressing cells (lanes *3*) and of cells expressing the truncated *cdc34-3* gene (lanes *4*) were assayed for ubiquitin-conjugating activity (Jentsch et al. 1987). (*A*) Assays for ubiquitin-histone H2B conjugation (uH2B) were performed either in the absence (lane *5*) or the presence of the E1 enzyme, and in the presence of ^{125}I-labeled ubiquitin (Ub) and histone H2B. (*B*) Ubiquitin-enzyme thiolester complexes were detected with added E1 (E1~u, lanes *1–3*), *CDC34* (CDC34~u, lane *1*), and *cdc34-3* (cdc34-3~u, lane *4*). (*C*) Same as in *B*, but samples were reduced prior to electrophoresis to cleave thiolester-mediated complexes selectively (for further details, see Jentsch et al. 1987).

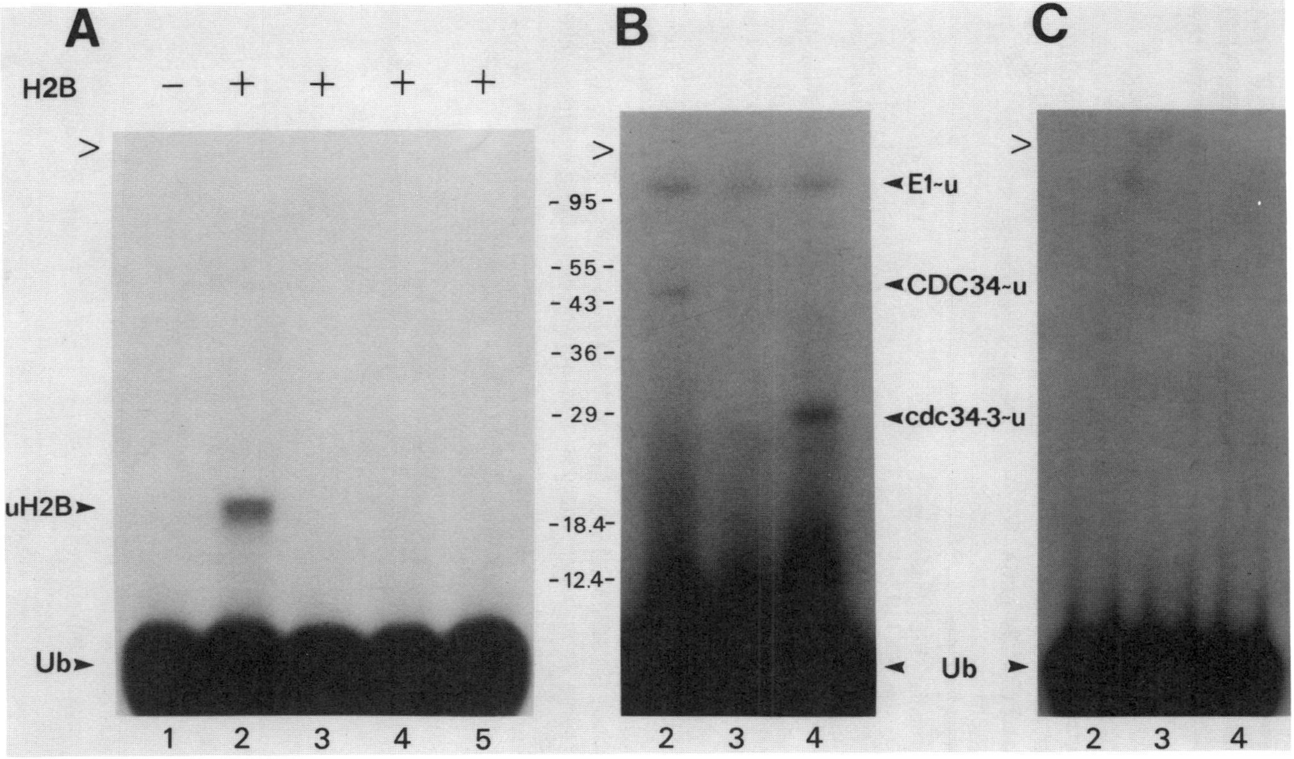

Figure 1 (*See facing page for legend.*)

zymes, these enzymes must differ in other properties to account for their distinct in vivo functions. The *CDC34(UBC3)* enzyme might catalyze ubiquitination of histones specifically in the G_1 phase of the cell cycle, in preparation for the initiation of DNA replication, whereas ubiquitination by the *RAD6(UBC2)* protein might be confined to the sites of damage in the chromosome.

ACKNOWLEDGMENTS

We thank M.G. Goebl, J. Yochem, and B. Byers for communicating their results prior to publication and for providing the cloned *CDC34* gene.

REFERENCES

Byers, B. 1981. Molecular genetics in yeast. *Proc. Alfred Benzon Symp.* **16:** 119.

Finley, D. and A. Varshavsky. 1985. The ubiquitin system: Functions and mechanisms. *Trends Biochem. Sci.* **10:** 343.

Friedberg, E.C., ed. 1983. *DNA repair*, p. 506. Freeman, San Francisco.

Hershko, A. and A. Ciechanover. 1986. The ubiquitin pathway for the degradation of intracellular proteins. *Prog. Nucleic Acid Res.* **33:** 19.

Hershko, A., H. Heller, S. Elias, and A. Ciechanover. 1983. Components of the ubiquitin-protein ligase system. *J. Biol. Chem.* **258:** 8206.

Jentsch, S., J.P. McGrath, and A. Varshavsky. 1987. The yeast DNA repair gene *RAD6* encodes a ubiquitin-conjugating enzyme. *Nature* **329:** 131.

Pickart, C.M. and I.A. Rose. 1985. Functional heterogeneity of ubiquitin carrier proteins. *J. Biol. Chem.* **260:** 1573.

Identification of Components That Regulate the Cell Cycle in Early *Drosophila* Embryos

P.H. O'Farrell, B. Edgar, and C. Lehner

Department of Biochemistry and Biophysics, University of California
San Francisco, California 94143

Nature produces patterns of exquisite detail and beauty. Our laboratory is interested in the molecular mechanisms involved in the embryonic development of such patterns. In *Drosophila*, the development of structures in their appropriate positions is controlled by regulators expressed in precisely localized patterns (Hafen et al. 1984; Scott and O'Farrell 1986). Analyses of mutants have shown that if one of these regulators is defective, the spatial patterns of other regulators are altered (Carroll and Scott 1986; DiNardo and O'Farrell 1987). Accordingly, a network of cross-regulations appears to specify the intricate patterns of expression of these regulators. A number of the developmental regulators have been shown to contain a highly conserved protein sequence, the homeo domain (McGinnis et al. 1984; Scott and Weiner 1984). Because of its presence in many of the developmental regulators (12 out of 20 analyzed) and its conservation even in humans, the protein domain must be involved in interactions important to function.

We have shown that the homeo domain is capable of sequence-specific DNA binding (Desplan et al. 1985). Our laboratory and several others are trying to dissect the circuitry of the regulatory network and to test the mechanisms of action of the component regulators of this network. In these studies, the processes under scrutiny appear to be intricate interactions of transcriptional regulators.

The Problem of Spatially Specific Growth Control

In the formation of embryonic pattern, cell proliferation rates are controlled in spatially specific patterns (Hartenstein and Campos-Ortega 1985; V. Foe, pers. comm.). The mechanisms involved are clearly fundamental to development and cellular growth control. The genes controlling the embryonic pattern must specify the detailed spatial patterns of cell-cycle lengths through unknown routes. The same powerful genetic and molecular approaches that led to the identification of the regulators that specify the *Drosophila* embryonic pattern might now be used to define the route of their action to control growth rates. To begin such studies, we want to define factors governing cell-cycle progression. As discussed below, a mutant string (*stg*) provides genetic access to the regulators, whereas evolutionary conservation has allowed us to use homology to "jump" to *Drosophila* from systems in which cell-cycle regulation has been investigated more intensively.

Early Embryonic Cell Cycles

Early development in many animals involves a series of very rapid, metasynchronous cell cycles that lack detectable gap phases and support little transcriptional activity (McKnight and Miller 1976; Zalokar 1976; Zalokar and Erk 1976; Newport and Kirschner 1982; Edgar and Schubiger 1986). In *Drosophila* embryos, the first extended interphase occurs during the 14th division cycle. It is during this period that large amounts of zygotic transcription first occur and the previously syncytial nuclei are cellularized (McKnight and Miller 1976; Zalokar 1976; Edgar and Schubiger 1986). After this, cell division resumes asynchronously in an intricate, bilaterally symmetric pattern. Segmental periodicities in this pattern suggest that identified genes controlling patterning of the embryo may also control the spatial program of cell-cycle lengths (V. Foe, pers. comm.). Cells in the extraembryonic tissues never divide after cycle 14, whereas embryonic cell lineages undergo between one and as many as six further division cycles before larval differentiation (Hartenstein and Campos-Ortega 1985).

Experiments treating *Drosophila* embryos with inhibitors suggest that when the zygotic genome first assumes control of the cell cycle at cycle 14, the control is at the point of entry into mitosis. For example, eliminating zygotic gene expression by blocking either transcription (during cycles 1–14) causes arrest in G_2 of cycle 14 (B. Edgar, unpubl.). Thus, maternal components seem to drive the cell cycle through completion of the 14th S phase, and control by the zygotic genome appears to begin discretely at initiation of the 14th mitosis. Perhaps developmental regulators influence the length of the cell cycle by regulating, in precise spatial patterns, the level or timing of expression of the zygotically required cell-cycle component(s).

The *stg* Locus May Encode a Regulator of Cell-cycle Progression

If a zygote were to lack a gene that regulated entry into mitosis, one might expect it to traverse the first 13 mitoses by virtue of the maternal product of the gene and to arrest in G_2 of cycle 14 when zygotic gene expression was first required. Embryos homozygous for mutations at the *stg* locus have this phenotype; the first 13 mitoses occur normally, but mitosis 14 is never initiated (Y.N. Jan, pers. comm.; S. DiNardo and P.H. O'Farrell, unpubl.). Furthermore, quantitative fuchsin staining has shown that most of the interphase-arrested nuclei in *stg* mutants do not become polyploid (B. Edgar, unpubl.) and thus are indeed arrested prior to mitosis 14. Nevertheless, the cells formed during cycle 14 undergo remarkably normal patterns of segmentation, gene expression, gastrulation, and differentiation (Jurgens et al. 1984; S. DiNardo, unpubl.), suggesting that *stg* mutations specifically block mitotic functions without profoundly perturbing other processes. This lack of pleiotropy suggests that *stg* may code for a cell-cycle regulator, rather than a metabolic or structural element necessary for cell division. The eight existing alleles of *stg* were discovered in a screen that was nearly saturating for point mutations affecting the cuticle pattern, yet no other mutations with a similar phenotype were uncovered (Jurgens et al. 1984). This last observation does not imply that *stg* encodes the only function required to advance the cell into mitosis. Recall (see above) that inhibition of RNA synthesis shows that maternally expressed products suffice for the first 13 division cycles. It may well be that cell-cycle regulation involves a multistep process but that the maternal supply of all but one of the required gene products is sufficient for more than 13 cycles. Indeed, the genetics suggest that this is so and that the limiting component is that encoded by *stg*. Recent analysis of hypomorphic alleles of *stg* has shown that reduced levels of *stg* activity lead to delays of mitosis (B.A. Edgar and P.H. O'Farrell, unpubl.). This is consistent with a possibility that a spatially patterned regulation of *stg* expression or activity could be responsible for the observed spatial differences in cell-cycle times that begin with the 14th cell cycle.

Cloning a *Drosophila* Cyclin Homolog

The evolutionary conservation of cell-cycle regulatory functions suggests that homology might be used to identify cell-cycle regulatory genes in *Drosophila*. These might provide useful tools in our studies of the spatial programming of cell-cycle times. There are two obvious candidates for initiating a search for homologs, cyclin, and *CDC2* of *Saccharomyces pombe*. Our early efforts to find a *Drosophila* homology of the cyclin-coding sequences has been successful.

Aided by advice from T. Hunt, we designed a degenerate oligonucleotide probe on the basis of a segment of the cyclin sequence conserved between clam, sea urchin, and frog. This oligonucleotide detected (frequency between 10^{-3} and 10^{-4}) positive clones in an oocyte cDNA library (provided by L. Jan and B. Hay). Sequencing revealed an open reading frame with extensive homologies to the cyclins of other species. In situ hybridization to polytene chromosomes localized the gene to 68D/E. A deletion that removes the locus encoding the identified *Drosophila* cyclin homolog, as well as flanking genes, is embryonic lethal but has no reported visible morphological defect. Because of the expected maternal contribution of cyclin RNA to the embryo, extensive zygotic development is not surprising.

Prospects

Molecular cloning and genetic tools can be used to test the hypothesis that cyclin functions as a level-dependent chronometer for cell-cycle progression and that it induces entry into mitosis at a particular threshold. Additionally, we intend to clone the *stg* locus and investigate its role, perhaps in setting the threshold for entry into mitosis. Ultimately, we hope to gain some understanding of how the regulators of the spatial pattern in the *Drosophila* embryo come to influence the cell cycle.

REFERENCES

Carroll, S.B. and M.P. Scott. 1986. Zygotically active genes that affect the spatial expression of the *fushi tarazu* segmentation gene during early *Drosophila* embryogenesis. *Cell* **45:** 113.

Desplan, C., J. Theis, and P.H. O'Farrell. 1985. The *Drosophila* developmental gene *engrailed* encodes a sequence specific DNA binding activity. *Nature* **318:** 630.

DiNardo, S. and P.H. O'Farrell. 1987. Establishment and refinement of segmental pattern in the *Drosophila* embryo: Spatial control of *engrailed* expression by pair–rule genes. *Genes Dev.* **1:** 1212.

Edgar, B.A. and G. Schubiger. 1986. Parameters controlling transcriptional activation during early *Drosophila* development. *Cell* **44:** 871.

Hafen, E., A. Kuroiwa, and W.J. Gehring. 19848. Spatial distribution of transcripts from the segmentation gene *fushi tarazu* during *Drosophila* embryonic development. *Cell* **37:** 833.

Hartenstein, V. and A. Campos-Ortega. 1985. Fate-mapping in wild-type *Drosophila melanogaster* I. The spatio-temporal pattern of embryonic cell divisions. *Wilhelm Roux's Arch. Dev. Biol.* **194:** 181.

Jurgens, G., E. Wieschaus, C. Nusslein-Volhard, and M. Kluding. 1984. Mutations affecting the pattern of the larval cuticle in *Drosophila melanogaster*. II. Zygotic loci on the third chromosome. *Wilhelm Roux's Arch. Dev. Biol.* **193:** 283.

McGinnis, W., R.L. Garber, J. Wirz, A. Kuroiwa, and W.J. Gehring. 1984. A homologous protein-coding sequence in *Drosophila* homeotic genes of the *Drosophila, Antennapedia*, and bithorax complexes. *Cell* **37:** 403.

McKnight, S.L. and D.L. Miller, Jr. 1976. Ultrastructural patterns of RNA synthesis during early embryogenesis in *Drosophila melanogaster*. *Cell* **8:** 305.

Newport, J. and M. Kirschner. 1982. A major developmental transition in early *Xenopus* embryos: I. Characterization and timing of cellular changes at the midblastula stage. *Cell* **30:** 675.

Scott, M.P. and P.H. O'Farrell. 1986. Spatial programming of gene expression in early *Drosophila* embryogenesis. *Annu. Rev. Cell Biol.* **2:** 49.

Scott, M.P. and A.J. Weiner. 1984. Structural relationships among genes that control development: Sequence homology between the *Antennapedia Ultrabithorax* and *fushi tarazu* loci of *Drosophila*. *Proc. Natl. Acad. Sci.* **81:** 4115.

Zalokar, M. 1976. Autoradiographic study of protein and RNA formation during early development of *Drosophila* eggs. *Dev. Biol.* **49:** 425.

Zalokar, M. and I. Erk. 1976. Division and migration of nuclei during embryogenesis of *Drosophila melanogaster*. *J. Microsci. Biol. Cell* **25:** 97.

Mitosis in *Drosophila*: The Centrosome Cycle in Early Embryogenesis

D.M. Glover, J. Axton, A. Cheshire, S. Delaney, M. Freeman, C. Girdham, C. Gonzalez, R. Karess, M. Leibowitz, S. Llamazares, S. Millar, D. O'Sullivan, J. Raff, R. Saunders, C. Sunkel, and W. Whitfield

Cancer Research Campaign, Eukaryotic Molecular Genetics Research Group, Department of Biochemistry
Imperial College of Science and Technology
London SW7 2AZ, England

Drosophila offers considerable advantages as a multicellular organism in which to study cell division. There are two major stages of *Drosophila* development at which one can expect to detect mutations affecting mitosis. The first is during early embryogenesis, when there are 13 nuclear divisions at about 10-minute intervals before individual nuclei undergo cellularization (see, e.g., Karr and Alberts 1986). During this time, there is little or no zygotic gene expression, and one might therefore expect a class of maternal-effect lethal mutations that disrupt the functions of the genes encoding these components. There are then only three to four more cell divisions during embryogenesis, and most larval development involves cell growth with the endoreduplication of DNA in the absence of mitosis. Imaginal cells, destined to form the adult organism and not necessary for the survival of the larva, divide throughout larval development as cells of the nervous system do. A second class of mitotic mutations can therefore be recognized in which homozygous mutant larvae survive, utilizing wild-type gene products supplied by their heterozygous mothers, but have imaginal cells that cannot proliferate. Therefore, death ensues during the late larval or early pupal stages, a phenotype first recognized in the analysis of repair-defective mutants (for review, see Gatti et al. 1983). We are working with late lethal mutations that affect chromosome condensation and segregation. In this paper, however, we concentrate on our analysis of maternal-effect mutations leading to aberrant mitoses during early embryogenesis.

The mutation *gnu* appears to affect nuclear division specifically during early embryogenesis. Homozygous *gnu* mothers produce embryos that develop giant nuclei (Freeman et al. 1986). We recently reported that fertilization of *GNU* eggs is not required for the giant nuclei to develop. This contrasts with the wild-type egg in which fertilization is required before any further development can take place (Freeman and Glover 1987). Whether or not the *GNU* egg is fertilized, any of the four products of female meiosis, the three polar bodies or female pronucleus, can participate in DNA synthesis to give giant nuclei. If the *GNU* egg is fertilized, the genomes of both parents replicate. It seems that the *GNU* mutation is overcoming the block to DNA synthesis that normally occurs following meiosis and is permitting DNA synthesis to occur in an uncontrolled manner that, in some way, prevents nuclear division. The gene therefore appears to play a role in the correct establishment of coordinated DNA replication and mitosis in zygotic development.

One of the striking features of *GNU* embryos is that although nuclear division does not take place, centrosomes continue to replicate. Normally, each wild-type interphase nucleus is associated with a single centrosome that replicates, giving rise to two daughter centrosomes. These migrate to opposite sides of the nucleus in prometaphase to nucleate the microtubules of a new spindle. The centrosomes of *GNU* embryos are dissociated from nuclei and do not function in the formation of mitotic spindles per se but are capable of nucleating asters of microtubles. They increase in number and migrate to the cortex of the developing *GNU* embryo as they would in the wild type, indicating that although the nuclear division cycle has been disrupted, the centrosome cycle continues independently. Centrosomal replication also occurs in unfertilized *GNU* eggs, indicating that all centrosomes can be maternally supplied. To determine whether centrosome replication is coupled to DNA replication, we have initially microinjected the DNA synthesis inhibitor, aphidicolin, into wild-type embryos. Under these circumstances, centrosome replication can continue for several rounds. We therefore suspect that the centrosome cycle will carry on independently of DNA replication in *GNU* embryos, although this has yet to be tested directly.

Mutations at four different loci on the third chromosome

result in a high proportion of circular mitotic figures in neuroblast cells, many of which are polyploid. This phenotype is seen with alleles of polo (Sunkel and Glover 1988), lodestar (C. Girdham et al., unpubl.), and merry-go-round (*mgr*) (Gonzalez et al. 1988). An allele of another mitotic mutant, aurora, generated by ethylmethane sulfonate (EMS), also displays this phenotype when made heterozygous with a chromosome deficient for the locus (M.H. Leibowitz and D.M. Glover, unpubl.). This mutation, originally identified because of its maternal effect, does not show mitotic phenotype in neuroblasts of homozygous larvae. Animals heterozygous for the overlapping deficiencies that define the cytological position of aurora die as pupae and have circular polyploid chromosome configurations. The integrity of microtubules appears to be required, as circular figures are no longer seen if the cells are treated with colchicine. Furthermore, circular figures are not seen in larvae homozygous for both *mgr* and *asp* (Ripoll et al. 1985), a gene required for the integrity of the spindle. One possible interpretation of this phenotype is that the chromosomes are in a circular arrangement around a monopolar spindle.

Chromosome nondisjunction is seen to occur during male meiosis with aurora and polo. The meiotic spindles of living cells are abnormal, and the spermatid nuclei are heterogeneous in size, indicative of a DNA content varying between as little as one quarter of the haploid DNA content up to as much as a tetraploid complement. *mgr* also shows meiotic abnormalities, in this case, resulting in the formation of tetraploid rather than haploid spermatid nuclei.

Females homozygous for the mutations lodestar, polo, and aurora survive to adulthood to lay eggs that die during embryogenesis. LODESTAR embryos begin to show a mitotic defect at about nuclear division cycle 9, in which there are multiply branched spindle structures with shared centrosomes. Weaker alleles show anaphase bridges leading to chromosome breakage. Immunocytological studies on POLO embryos also reveal highly branched mitotic spindles with broad irregular poles that show no distinct staining for a centrosome-associated antigen. The centrosome-associated antigen, Bx63, is present in these embryos not in discrete structures, but in particulate matter that coalesces throughout the abnormal development of the embryo. The mitotic spindles of AURORA embryos can also be highly branched. Furthermore, whereas the anaphase spindles are focused on well-defined "dot-like" centrosomes, the broad, telophase-like spindles appear to be nucleated from points around the nuclear envelope.

Although these and other aspects of the phenotypes suggest lesions affecting the centrosome, it is prudent to remain cautious as to the precise nature of the primary defects because aberrations in one step of the mitotic cycle could have epistatic effects.

The antigen recognized by the monoclonal antibody Bx63 is associated with both the centrosome and the nucleus; it accumulates during oogenesis in the oocyte nucleus and becomes cytoplasmic in the fertilized egg, and it is then assembled into centrosomes throughout cleavage. The antibody identifies two proteins, 185 and 66 kD, on Western blots (Frasch et al. 1986). We have used both the monoclonal antibody and polyclonal sera to screen λgt11 expression libraries and have cloned DNA that directs the synthesis of fusion proteins that share epitopes with the 185-kD centrosomal antigen (Whitfield et al. 1988). The DNA that hybridizes in situ to the salivary gland chromosome region 88E4-8 directs the synthesis of transcript present at all developmental stages, being most abundant in adult females and embryos. We are currently generating deficiencies in this chromosomal region so that we might ultimately carry out saturation mutagenesis to identify all complementation groups within the region.

Clearly, much work needs to be done to clarify the nature, relationships, and genetic interactions between the genes we are studying. The genetic approach will be invaluable in understanding mitosis, but it has limitations that can be overcome with the concerted application of molecular studies. We have strategies for cloning each of these genes, and it should only be a matter of time before they are sequenced and expressed in *Escherichia coli* so that the gene product can be used as an immunogen. Antibodies raised in this way will be powerful tools in analyzing the functions of these gene products in *Drosophila* cells.

REFERENCES

Frasch, M., D.M. Glover, and H. Saumweber. 1986. Nuclear antigens follow different pathways into daughter nuclei during mitosis in early *Drosophila* embryos. *J. Cell Sci.* **82:** 155.

Freeman, M. and D.M. Glover. 1987. The *gnu* mutation causes inappropriate DNA synthesis in unfertilized and fertilized eggs. *Genes Dev.* **1:** 924.

Freeman, M., C. Nüsslein-Volhard, and D.M. Glover. 1986. The dissociation of nuclear and centrosomal division in *gnu*, a mutation causing giant nuclei in *Drosophila. Cell* **46:** 457.

Gatti, M., S. Pimpinelli, C. Bove, B.S. Baker, D.A. Smith, A.T.C. Carpenter, and P. Ripoll. 1983. Genetic control of mitotic cell division in *Drosophila melanogaster. Proc. Int. Congr. Genet.* **2:** 193.

Gonzalez, C., J. Casal, and P. Ripoll. 1988. Functional monopolar spindles caused by the mutation in *mgr*, a cell division gene of *Drosophila melanogaster. J. Cell. Sci.* **89:** 39.

Karr, T.L. and B.M. Alberts. 1986. Organisation of the cytoskeleton in early *Drosophila* embryos. *J. Cell Biol.* **102:** 1494.

Ripoll, P., S. Pimpinelli, M.M. Valdivia, and J. Avila. 1985. A cell division mutant of *Drosophila* with a functionally abnormal spindle. *Cell* **41:** 907.

Sunkel, C.E. and D.M. Glover. 1988. *polo*: a mitotic mutant of *Drosophila* displaying abnormal spindle poles. *J. Cell. Sci.* **89:** 25.

Whitfield, W.G.F., S.E. Millar, H. Saumweber, M. Frasch, and D.M. Glover. 1988. Cloning of a gene encoding an antigen associated with the centrosome in *Drosophila. J. Cell. Sci.* **89:** 467.

Molecular Cloning of Human Genes That Complement the Temperature-sensitive Cell-cycle Mutants of BHK-21 Cells

T. Nishimoto, R. Kai, T. Sekiguchi, M. Ohtsubo, H. Sasaki, N. Furuno, S. Uchida, K. Miyauchi, and N. Nishitani

Department of Molecular Biology, Graduate School of Medical Science
Kyushu University, Fukuoka 812, Japan

The development of DNA-mediated gene transfer (DMGT) has changed the biology of somatic cells. Many phenotypes of somatic cells can now be discussed on the gene level.

Nishimoto and Basilico (1978) established an efficient method for selecting temperature-sensitive mutants for cell growth, isolated 219 such mutants from BHK cells, and classified them into 18 complementation groups (Nishimoto et al. 1981). These mutants, including eight G_1 mutants, one possible S mutant, and two G_2 mutants, are being used to investigate the molecular mechanism of the cell cycle. As a first step, we have been cloning the human genes that complement temperature-sensitive mutations. Thus far, we have cloned the two human genes, *CCG1* and *RCC1*.

Human *CCG1* Gene

The human *CCG1* gene was cloned from the total DNA of KB cells by DMGT, using the *ts*BN462 cells as a recipient of ts^+ transformation (Sekiguchi et al. 1987, 1988).

Character of tsBN462 *Cells*

*ts*BN462 cells cannot progress into the S phase at 39.5°C, following release from isoleucine deprivation. Their progression into the S, G_2, and M phases seems to be normal. The execution point is about 2 hours prior to the S phase, but total RNA synthesis is reduced slightly at 39.5°C from early G_1 phase (T. Sekiguchi et al., unpubl.). Upon fusion between *ts*BN462H (*ts*, $HPRT^-$) and B1 (ts^+, TK^-) cells, the temperature-sensitive phenotype cosegregates with the HPRT$^-$ phenotype, suggesting that the *ts*BN462 mutation is located on the X chromosome (Nishimoto et al. 1982). Consistently, many temperature-sensitive mutants isolated independently, e.g., *ts*13 cells (Talavera and Basilico 1977), were classified into the same complementation group as *ts*BN462 cells.

Isolation of the Human CCG1 *Gene*

For the isolation of genomic DNA, high-molecular-weight DNA of the KB cell was transfected into the *ts*BN462 cell by the calcium phosphate precipitation method; the conserved human DNA through the primary, secondary, and tertiary *ts*$^+$ transformation was cloned using the cosmid vector pHC79 (Sekiguchi et al. 1987). About 70 kb of human DNA, which is flanked by the hamster DNA and is covered by seven cosmid clones, was isolated. The isolated human DNA originates from the human X chromosome and seems to locate on the p^{ter}-q^{21} region.

Because expression of the *CCG1* gene is very low, we could not detect any cDNA clones in the Okayama-Berg cDNA library. Thus, the cDNA library was constructed from the secondary *ts*$^+$ transformant of the *ts*BN462 cells in which the transfected human *CCG1* gene is amplified. The isolated cDNA (5.3 kb) complements both the *ts*BN462 and *ts*13 cells (Sekiguchi et al. 1988), consistent with the fact that these mutants belong to the same complementation group.

Putative Gene Product of the CCG1 *Gene*

The isolated cDNA has an open reading frame of 4662 bp, encoding a 178,768-dalton protein. The putative *CCG1* protein, which has no homology with proteins in the NBRF data bank, is hydrophilic, with a tandem repeat of 120 amino acids in the carboxy-terminal region; an amino acid sequence (PPKKKRRV)

Figure 1 Structure of the putative *CCG1* protein. (*a*) Tandem repeated region of 120 amino acids. (Closed circles) Same amino acids; (open circles) biologically similar amino acids. (*b*) Comparison of the putative nuclear location signal of human *CCG1* protein with that of the SV40 T antigen. (*C*) Schematic structure of the putative human *CCG1* protein. Tandem repeated regions are indicated by hatching and small dots; the large centered dot shows the position of the nuclear location signal. The whole sequence of the human *CCG1* gene will appear in the EMBL/GenBank/DDBJ Nucleotide Sequence Data Bases, under the accession number X07024, *CCG1*.

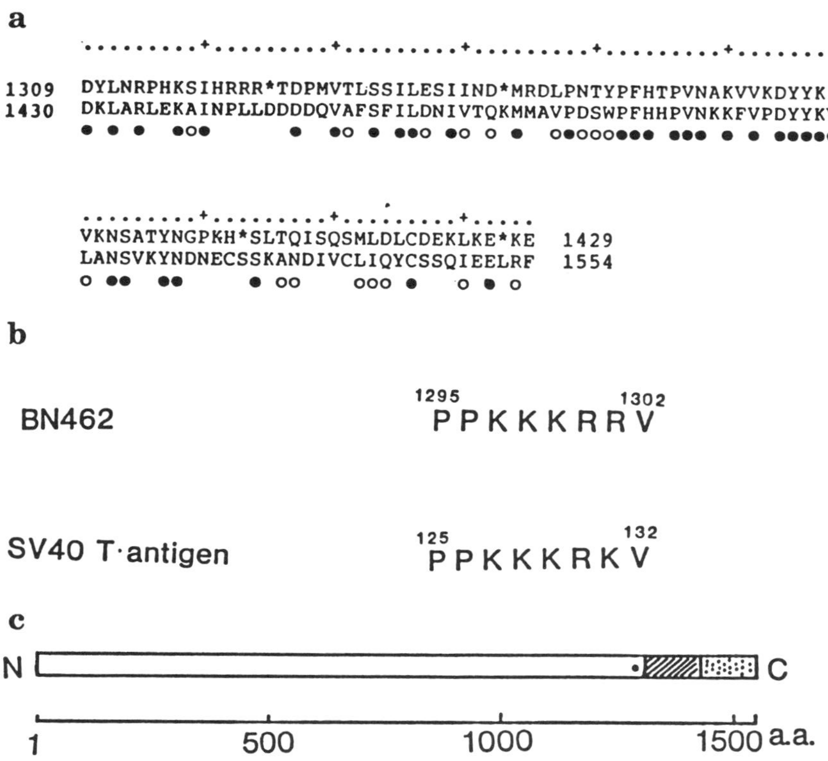

Figure 1 *(See facing page for legend.)*

similar to the nuclear location signal sequence of the SV40 large T antigen is located immediately before the tandem repeat of amino acids (Fig. 1).

RCC1 Gene

The human *RCC1* gene complements the *ts*BN2 mutation (Kai et al. 1986; Ohtsubo et al. 1987), which shows premature chromosome condensation (PCC) at 39.5°C (Nishimoto et al. 1978). Because all mitotic specific events identified so far can be induced in *ts*BN2 cells at 39.5°C, we considered that *ts*BN2 cells may have a temperature-sensitive defect in the regulatory mechanism for the onset of chromosome condensation (Nishimoto et al. 1981; Ajiro et al. 1983; Yamashita et al. 1983, 1985).

Structure of the Human RCC1 *Gene*

The total base sequence of the cosmid clone B95 (45 kb) was determined, and it was found that the *RCC1* gene has two promoter regions, one of which is located outside of this cosmid clone (N. Furuno et al., in prep.). The *RCC1* gene has 11 exons, and the repeated domain of amino acids in the putative *RCC1* protein (Ohtsubo et al. 1987) consists of 8 exons. Each repeat seems to correspond to a single exon, except the middle one of the repeated domain.

Identification of the RCC1 *Gene Product*

Antibodies against the putative *RCC1* gene product were prepared, using as antigens a peptide of 15–16 amino acids constructed according to the base sequence of either the 5′ region immediately before or at the repeated domain. Only one antibody against the peptide located immediately before the amino-acid-repeated domain recognized a 45-kD protein, the size of the putative *RCC1* gene product. This 45-kD protein is present in HeLa cells and in the ts^{+} transformant of *ts*BN2 cells transfected with the human *RCC1* cDNA but not in the *ts*BN2 cells.

The *RCC1* protein is located mainly in the nucleus. The *RCC1* protein of HeLa cells adsorbed to a DNA-cellulose column and was eluted with 0.3 M NaCl, suggesting that the *RCC1* protein can bind to DNA. The *RCC1* protein is present throughout the cell cycle and does not disappear at a specific phase of the cell cycle.

Cloning of Hamster RCC1 *DNA*

To determine the site of the *ts*BN2 mutation and also to exclude the possibility that the *RCC1* gene is a suppressor, we are cloning the *RCC1* cDNA from both BHK-21 and *ts*BN2 cells. Thus far, the *RCC1* cDNA of BHK-21 cells has been cloned. This cDNA can complement the *ts*BN2 mutation and is similar to the human *RCC1* cDNA by as much as 90%, calculated by base sequence data (S. Uchida et al., unpubl.).

Proposed Model of the Regulatory Mechanism for the Onset of Chromosome Condensation

Schlegel et al. (1986) found that PCC can be induced in BHK cells in the presence of hydroxyurea and caffeine. Because this phenomenon occurs at both 33.5°C and 39.5°C, we investigated the ability of temperature-sensitive mutants to induce PCC at 39.5°C with caffeine and hydroxyurea (Table 1). In three mutants, *ts*BN75, *ts*BN250, and *ts*BTN1, PCC induction was reduced at 39.5°C. One of these mutants, the *ts*BN75 cell, has the same defect as the *ts*85 cell (N. Ohtsubo and T. Nishimoto, in prep.). The *ts*85 cell is a mutant of the ubiquitin-activating enzyme (Finley and Varshavsky 1985), suggesting the possible involvement of ubiquitin for the onset of chromosome condensation.

Table 1 PCC Induction by Caffeine in Temperature-sensitive Mutants Synchronized at Early S Phase by Hydroxyurea

	33.5°C		40°C (preheat 4 hr)	
Cell lines	3 hr (%)	5 hr (%)	3 hr (%)	5 hr (%)
Wild-type BHK	45.6	45.0	48.9	25.5
*ts*BTN1	6.5	5.4	1.7	0.76
*ts*BTN1^R	6.5	4.9	6.4	6.3
*ts*BN2	24.5	44.4	—	—
*ts*BN51	28.1	27.5	33.7	18 .8
*ts*BN63	11.9	13.5	10.2	6.3
*ts*BN67	4.9	8.2	13.4	6.4
*ts*BN75	26.7	29.2	5.7	2.1
*ts*BN74^R	20.0	38.4	51.1	50.6
*ts*BN250	8.9	13.8	0.0	0.0
*ts*BN250^R-1	22.8	20.4	19.5	12.7
*ts*BN250^R-2	11.1	15.5	14.3	21.2
*ts*BN462	28.9	30.7	20.2	12.2

Following release from isoleucine deprivation at 33.5°C, cells were synchronized by hydroxyurea (2.5 mM), incubated at 33.5°C or 39.5°C for 4 hr, and caffeine was added. *ts*BTN1^R, *ts*BN75^R, *ts*BN250^R are spontaneous revertants of *ts*BTN1, *ts*BN75, and *ts*BN250 cells, respectively.

We assumed that the *RCC1* protein is a kind of repressor for the expression of a gene coding either a trigger protein of chromosome condensation or the maturation-promoting factor (MPF) and is inactivated specifically at the start of the G_2 phase. Because two independently isolated G_2 mutants have the same defect, i.e., the temperature-sensitive ubiquitin-activating enzyme, this enzyme should have a key function in the start of the G_2–M phase. The ubiquitin-activating enzyme is the first one required to conjugate ubiquitin to the target protein, and ubiquitin is involved in either the inactivation or the degradation of proteins. Therefore, we assumed that the *RCC1* protein is inactivated by the ubiquitin system. If ubiquitin is required for the onset of chromosome condensation, at least one more protein in addition to the ubiquitin-activating enzyme (the ubiquitin-conjugating enzyme) is also needed for this process (Finley and Varshavsky 1985). It will be interesting to see whether a temperature-sensitive mutant of the ubiquitin-conjugating enzyme also shows the G_2 block.

DISCUSSION

We are far from discussing all of the molecular events of the cell cycle. As a first step, we think it is essential to accumulate the cloned cell-cycle genes, although the function of cloned genes is unknown in many cases. Our present results show that cloning of the human gene, using temperature-sensitive hamster mutants as a recipient of DMGT, is practical. We are now cloning the human genes complementing the temperature-sensitive mutants *ts*BN75, *ts*BTN1, and tsBN250, which are suggested to be involved in the induction of PCC. In the future, it will be important to isolate more temperature-sensitive cell-cycle mutants in order to clone many human genes.

REFERENCES

Ajiro, K., T. Nishimoto, and T. Takahashi. 1983. Histone H1 and H3 phosphorylation during premature chromosome condensation in a temperature sensitive mutant (*ts*BN2) of baby hamster kidney cells. *J. Biol. Chem.* **258:** 4534.

Finley, D. and A. Varshavsky. 1985. The ubiquitin system: Functions and mechanisms. *Trends Biochem. Sci.* **10:** 343.

Kai, R., M. Ohtsubo, M. Sekiguchi, and T. Nishimoto. 1986. Molecular cloning of a human gene that regulates chromosome condensation and is essential for cell proliferation. *Mol. Cell. Biol.* **6:** 2027.

Nishimoto, T. and C. Basilico. 1978. Analysis of a method for selecting temperature sensitive mutants of BHK cells. *Somatic Cell Genet.* **4:** 323.

Nishimoto, T., E. Eilen, and C. Basilico. 1978. Premature chromosome condensation in a tsDNA$^-$ mutant of BHK cells. *Cell* **15:** 475.
Nishimoto, T., R. Ishida, K. Ajiro, S. Yamamoto, and T. Takashi. 1981. The synthesis of protein(s) for chromosome condensation may be regulated by post-transcriptional mechanism. *J. Cell. Physiol.* **109:** 299.
Nishimoto, T., T. Sekiguchi, R. Kai, K. Yamashita, T. Takahashi, and M. Sekiguchi. 1982. Large-scale selection and analysis of temperature-sensitive mutants for cell reproduction from BHK cells. *Somatic Cell Genet.* **6:** 811.
Ohtsubo, M., R. Kai, N. Furuno, T. Sekiguchi, M. Sekiguchi, H. Hayashida, K. Kuma, T. Miyata, S. Fukushige, T. Murotsu, K. Matsubara, and T. Nishimoto. 1987. Isolation and characterization of the active cDNA of the human cell cycle gene (RCC1) involved in the regulation of onset of chromosome condensation. *Genes Dev.* **1:** 585.
Schlegel, R. and A.B. Pardee. 1986. Caffeine-induced uncoupling of mitosis from the completion of DNA replication in mammalian cells. *Science* **232:** 1264.
Sekiguchi, T., T. Miyata, and T. Nishimoto. 1988. Molecular cloning of the cDNA of human X chromosomal gene (CCG1) which complements the temperature sensitive G_1 mutants, *ts*BN462 and *ts*13, of teh BHK cell line. *EMBO J.* **7:** (in press).
Sekiguchi, T., M.C. Yoshida, M. Sekiguchi, and T. Nishimoto. 1987. Isolation of a human X chromosome-linked gene essential for progression from G1 to S phase of the cell cycle. *Exp. Cell. Res.* **169:** 395.
Talavera, A. and C. Basilico. 1977. Temperature sensitive mutants of BHK cells affected in cell cycle progression. *J. Cell. Physiol.* **92:** 425.
Yamashita, K., T. Nishimoto, and M. Sekiguchi. 1983. Analysis of protein associated with chromosome condensation in baby hamster kidney cells. *J. Biol. Chem.* **258:** 4534.
Yamashita, K., F.M. Dais, P.N. Rao, M. Sekiguchi, and T. Nishimoto. 1985. Phosphorylation of nonhistone proteins during premature chromosome condensation in a temperature-sensitive mutant, tsBN2. *Cell Struct. Funct.* **10:** 259.

DNA Topoisomerase II: Facilitated Degradation Accompanies Chromosome Decondensation during Early G_1

W.C. Earnshaw and M.M.S. Heck[1]

Department of Cell Biology and Anatomy, Johns Hopkins University School of Medicine, Baltimore, Maryland 21205

The original conception of the cell cycle was based on the highly visible events of mitosis, which center around the dramatic structural reorganizations of the nucleus and chromosomes. Early cytologists thought the cell was most active during mitosis and was quiescent during the long interphase period. With the recognition that DNA replication occupies a discrete period during interphase, we now typically divide the cycle into four phases: M, G_1, S, and G_2 (mitosis, first gap, DNA synthesis, and second gap).

It is perhaps surprising that relatively few major cell-cycle control points have been recognized to date. These include the point of commitment to enter into a cycle of DNA replication and division, the restriction (Pardee et al. 1978) or Start (Hartwell et al. 1974) point, the entry into S phase, and the entry into mitosis. In this paper, we are concerned with the structural transformation of the chromosomes that occurs during mitosis, particularly the way in which this transformation appears to be controlled. We present here experiments that lead us to postulate the existence of a new cell-cycle control point located at the transition from mitosis to the subsequent G_1 phase.

Mitotic chromosomes are composed of approximately equal masses of DNA, histone, and nonhistone proteins. In 1977, Laemmli and his co-workers (Adolph et al. 1977) obtained ex-

[1]Present address: Department of Embryology, Carnegie Institute of Washington, Baltimore, Maryland 21210.

perimental evidence confirming and greatly extending previous models which suggested that the structure of the mitotic chromosomes might be organized by interactions among the nonhistone proteins (Dounce et al. 1973). Laemmli proposed that the structural nonhistone proteins were found in a discrete substructure, the chromosome scaffold, located centrally in each chromatid, and that the DNA radiated outward from this scaffold in loop domains (Laemmli et al. 1978).

After an initial broad acceptance, the scaffold hypothesis came under fire, primarily because the nonhistone substructure could not be visualized directly by electron microscopy in thin sections of chromosomes. We therefore set out to obtain specific antibody probes that would recognize scaffold components in intact chromosomes and permit us to ask whether these components would be concentrated at the base of the chromatin loops, as predicted by Laemmli et al. (1978).

The first scaffold protein for which a good antibody was available was the 170-kD major scaffold protein, previously termed Sc1 (Lewis and Laemmli 1982). In collaboration with the laboratory of L. Liu, we showed that this protein is the well-known enzyme DNA topoisomerase II (Earnshaw et al. 1985). P. Fisher and co-workers independently found that topoisomerase II is also a prominent component of the interphase nuclear matrix (Berrios et al. 1985). These results indicated that topoisomerase II is not just a soluble enzyme, but rather, appears to show tight associations with other nuclear proteins. In fact, topoisomerase II may turn out to be an enzyme like dynein and myosin, whose activity is regulated by relatively stable structural interactions.

We have used a combination of light and electron microscopy experiments to map the distribution of topoisomerase II in mitotic chromosomes. Both methods indicate that the enzyme is not found distributed widely throughout the mitotic chromatin but, rather, that it appears to occupy a restricted location at the base of the radial loop domains (Earnshaw and Heck 1985). The immunoelectron microscopy experiments indicated further that the topoisomerase II appears to be distributed in a great many discrete foci scattered throughout the arms of the chromosomes (which were, of necessity, somewhat expanded to permit immunolabeling). We think it likely that these foci are anchoring complexes, each of which is responsible for topological closure of a number of chromosomal loop domains. The existence of such complexes surrounded by rosettes of loops had been suggested previously by Comings (1977). It will remain for future experiments to discern the significance and composition of the putative anchoring complexes.

The most obvious structural attribute of mitotic chromosomes is that they look extremely different from the interphase nucleus. This is due, in part, to changes in cell physiology, (e.g., there is no ongoing transcription during mitosis), but it primarily reflects a dramatic reorganization of the highest orders of packaging of the chromatin. Such a reorganization might be accomplished in several ways. For example, posttranslational modifications of the subunits might cause a dramatic change in their tendency to associate into macromolecular aggregates. This is apparently the case for the nuclear lamina (which disassembles into soluble subunits during mitosis [Gerace and Blobel 1980]). Alternatively, the reorganization of the chromatin that accompanies mitosis might reflect the synthesis of stage-specific proteins that are present during mitosis and are degraded as chromosomes decondense. Here, we present evidence that topoisomerase II is such a protein.

We have focused our studies on the role of topoisomerase II in mitosis for three reasons. First, as mentioned above, the enzyme appears to comprise a significant fraction of the mass of the chromosome scaffold. Second, the activity of topoisomerase II is required for chromosomes to disjoin successfully at mitotic anaphase (Sundin and Varshavsky 1981; Holm et al. 1985; Uemura and Yanagida 1986). Third, both genetic and biochemical experiments indicate that topoisomerase II seems to be required for at least the terminal stages of mitotic chromosome condensation (Uemura et al. 1987; unpublished experiments cited in Newport and Spann 1987).

The earliest evidence that topoisomerase II expression varies across the cell cycle came from immunofluorescence experiments designed to examine the intracellular location of the enzyme. As expected, the enzyme was found only in interphase nuclei and in mitotic chromosomes; but surprisingly, many of the interphase nuclei were found to express low or undetectable amounts of the antigen (Earnshaw et al. 1985). Indirect immunofluorescence analysis of cells sorted with respect to DNA content (position in the cell cycle) by flow cytometry indicated that all of the negative cells had a prerreplicative DNA content and were thus apparently in G_1 phase.

Two obvious possibilities were raised by this result. First, the cellular content of topoisomerase II might vary dramatically across the cell cycle. Second, the antigen-negative cells might not be proliferating any longer, and topoisomerase II might be present only in proliferating cells.

We have shown by analysis of two embryonic systems, in which proliferation ceases under preprogrammed developmental control, that the cellular content of topoisomerase II is strictly correlated with the proliferative state of the cell (Heck and Earnshaw 1986). When either erythroblasts or myoblasts cease dividing, they lose all detectable topoisomerase II rapidly. In the former case, the decrease in antigen level is at least 500-fold. Furthermore, we also showed that when exposure to lectins induced quiescent peripheral blood lymphocytes to commence proliferation, topoisomerase II was synthesized de novo at the onset of S phase (Heck and Earnshaw 1986). A link between cellular topoisomerase II enzymatic activity and proliferative state has also been found by other workers (Duguet et al. 1983; Sullivan et al. 1986).

More recently, we have shown that in addition to a connection between topoisomerase II levels and proliferative state, the levels of the enzyme also fluctuate significantly across the cell cycle. This was done by examination of the total level of topoisomerase II antigen at a number of points across the cell cycle, using the method of centrifugal elutriation to fractionate populations of cells in exponential growth. As controls, we have also followed the levels of DNA topoisomerase I and a second major mitotic chromosome scaffold protein Sc2 (135 kD). In these experiments, we subjected the total cellular proteins from an equivalent number of cells corresponding to each elutriated fraction to SDS-PAGE and examined antigen levels by quantitative immunoblotting. In parallel, identical samples were electrophoresed in a similar gel and stained with Coomassie blue, and the dye was subsequently eluted to give a measure of the total protein content of each sample. We could thus determine the ratio of each antigen to the total amount of protein present at each point of the cell cycle. These results are presented in Table 1.

The data in Table 1 show that the expression of topoisomerase II is significantly different from that of topoisomerase I. The levels of the latter correlate closely with the overall level of cellular protein, whereas the level of topoisomerase II is significantly lower in early G_1 than at later times of the cycle. Thus, although the bulk of cellular proteins

Table 1 Relative Levels of Topoisomerase I and II across the Cell Cycle

Cell-cycle phase	Amount of specific antigen/total protein content	
	topoisomerase I	topoisomerase II
Early G_1	1.0[a]	1.0[a]
Mid-S	1.06 ± 0.03	1.43 ± 0.33
Late G_2/M	1.0 ± 0.11	1.69 ± 0.44

[a]These values were normalized to 1.0 to facilitate comparison.

are halved by the process of mitosis, the level of topoisomerase II apparently decreases by more than twofold, either during or immediately subsequent to mitosis. This implies that the stability of DNA topoisomerase II may be significantly less in early G_1 than at other points of the cycle. We have shown that this is, in fact, the case. The half-life of topoisomerase II in an exponentially growing cell culture is 12 ± 3 hours (Heck and Earnshaw 1988). This value decreases dramatically during the transition from mitosis to G_1. At this time, the half-life drops to 1.8 ± 0.4 hours, a decrease of approximately sevenfold (Fig. 1). Thus, the cycle of chromatin condensation/decondensation at mitosis is accompanied by a parallel cycle of synthesis/degradation of topoisomerase II. It is not currently known whether this degradation of topoisomerase II is a cause or a consequence of the process of chromosome decondensation.

Topoisomerase II is not alone in showing such alterations in its stability in the time period around mitosis, particularly at or subsequent to the onset of anaphase. Other proteins that undergo similar changes in stability include the cyclins from both sea urchin (Evans et al. 1983) and clam (Swenson et al. 1986), as well as thymidine kinase (J.L. Sherley and T.J. Kelly, in prep.). It is likely that numerous other proteins will be found to also undergo programmed degradation during the exit from mitosis. Thus, we propose that the transition from mitosis into the subsequent G_1 phase may be a significant point of cell-cycle control.

How might such selective degradation occur at this particular stage of the cell cycle? Two factors might be significant. First is the pulsed release of intracellular calcium that has been observed during anaphase (Poenie et al. 1986). This could activate cytosolic calcium-dependent proteases, one of which (calpain II) has recently been shown to become associated with the condensed chromosomes at mitosis (Schollmeyer 1988).

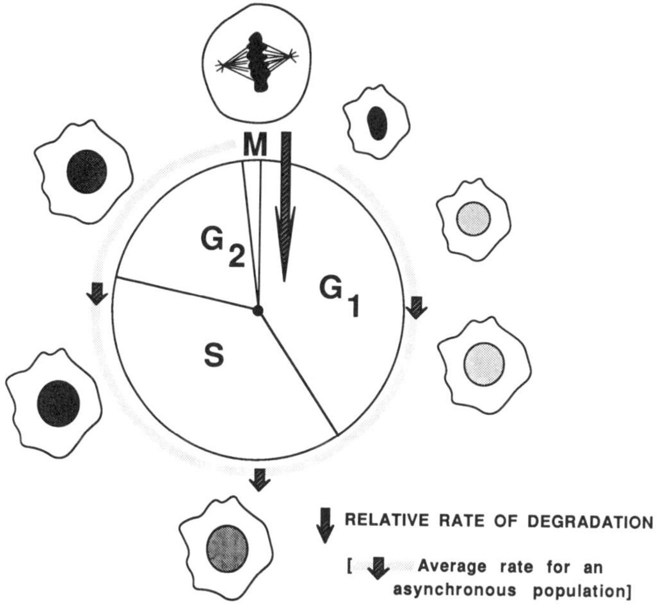

Figure 1 Stability of topoisomerase II across the cell cycle. Diagram summarizing measurements of the half-life of DNA topoisomerase II in exponentially growing populations and during the transition from mitosis to early G_1.

Second, during mitosis, the barrier between the nucleus and cytoplasm is removed. Thus, nuclear proteins may be accessible to cytosolic proteases that are normally excluded from the nucleus during the bulk of the cycle. This accessibility is expected to be maximal from prometaphase-anaphase, but it may continue during telophase and early G_1 as well. At these later times, the chromatin remains highly condensed so that even though a nuclear envelope now exists, it is likely that even proteins with nuclear targeting signals may remain trapped in the cytoplasm due to insufficient room within the forming nucleus.

Why would such a programmed degradation occur? As suggested by earlier studies of the cyclins (Evans et al. 1983; Swenson et al. 1986), the mitosis-G_1 transition might be a time of "resetting" of the cell-cycle program. Removal of certain key proteins involved in replication and mitosis might provide the cell with the option to choose between further proliferation and either hesitation or entry into G_0 when the Start point is subsequently reached. In this regard, it may be significant that the stability of topoisomerase II in nontransformed cells is three- to fourfold less than that in transformed cells (Heck et al. 1988). It is tempting to speculate that a more efficient operation of the resetting point may be one factor responsible for the greater tendency of nontransformed cells to cease dividing and embark on pathways of terminal differentiation.

ACKNOWLEDGMENTS

We thank Carol Cooke for her invaluable help during many stages of this project. We also acknowledge our collaborators in various stages of this work, including Walter Hittelman, Leroy Liu, and Brian Halligan. These studies were supported by National Institutes of Health grant GM-30985 to W.C.E. M.M.S.H. is a graduate fellow of the National Science Foundation.

REFERENCES

Adolph, K.W., S.M. Cheng, J.R. Paulson, and U.K. Laemmli. 1977. Isolation of a protein scaffold from mitotic HeLa cell chromosomes. *Proc. Natl. Acad. Sci.* **11:** 4937.

Berrios, M., O. Osheroff, and P.A. Fisher. 1985. In situ localization of DNA topoisomerase II, a major polypeptide component of the *Drosophila* nuclear matrix fraction. *Proc. Natl. Acad. Sci.* **82:** 4142.

Cornings, D.E. 1977. Mammalian chromosome structure. *Chromosomes Today* **6:** 19.

Dounce, A.L., S.K. Chanda, and P.L. Townes. 1973. The structure of higher eukaryotic chromosomes. *J. Theor. Biol.* **42:** 275.

Duguet, M., C. Lavenot, F. Harper, G. Mirambeau, and A.-M. De Recondo. 1983. DNA topoisomerases from rat liver: Physiological variations. *Nucleic Acids Res.* **11:** 1059.

Earnshaw, W.C. and M.M.S. Heck. 1985. Localization of topoisomerase II in mitotic chromosomes. *J. Cell Biol.* **100:** 1716.

Earnshaw, W.C., B. Halligan, C.A. Cooke, M.M.S. Heck, and L.F. Liu. 1985. Topoisomerase II is a structural component of mitotic chromosome scaffolds. *J. Cell Biol.* **100:** 1706.

Evans, T., E.T. Rosenthal, J. Youngblom, D. Distel, and T. Hunt. 1983. Cyclin: A protein specified by maternal mRNA in sea urchin eggs that is destroyed at each cleavage division. *Cell* **33:** 389.

Gerace, L. and G. Blobel. 1980. The nuclear envelope lamina is reversibly depolymerized during mitosis. *Cell* **19:** 277.

Hartwell, L.H., J. Culotti, J.R. Pringle, and B.J. Reid. 1974. Genetic control of the cell division cycle in yeasts. *Science* **183:** 46.

Heck, M.M.S. and W.C. Earnshaw. 1986. Topoisomerase II: A specific marker for cell proliferation. *J. Cell Biol.* **103:** 2569.

Heck, M.M.S., W.N. Hittelman, and W.C. Earnshaw. 1988. Differential regulation of DNA topoisomerase I and II during the eukaryotic cell cycle. *Proc. Natl. Acad. Sci.* **85:** 1086.

Holm, C., T. Goto, J.C. Wang, and D. Botstein. 1985. DNA topoisomerase II is required at the time of mitosis in yeast. *Cell* **41:** 553.

Laemmli, U.K., S.M. Cheng, K.W. Adolph, J.R. Paulson, J.A. Brown, and W.R. Baumbach. 1978. Metaphase chromosome structure: The role of nonhistone proteins. *Cold Spring Harbor Symp. Quant. Biol.* **42:** 351.

Lewis, C.D. and U.K. Laemmli. 1982. Higher order metaphase chromosome structure: Evidence for metalloprotein interactions. *Cell* **17:** 849.

Newport, J. and T. Spann. 1987. Disassembly of the nucleus in mitotic extracts: Membrane vesicularization, lamina disassembly, and chromosome condensation are independent processes. *Cell* **48:** 219.

Pardee, A.B., R. Dubrow, J.L. Hamlin, and R.F. Kletzien. 1978. Animal cell cycle. *Annu. Rev. Biochem.* **47:** 715.

Poenie, M., J. Alderton, R. Steinhardt, and R. Tsien. 1986. Calcium rises abruptly and briefly throughout the cell at the onset of anaphase. *Science* **233:** 886.

Schollmeyer, J.E. 1988. Calpain II involvement in mitosis. *J. Cell Biol.* (in press).

Sullivan, D.M., B.S. Glisson, P.K. Hodges, S. Smallwood-Kentro, and W.E. Ross. 1986. Proliferation dependence of topoisomerase II mediated drug action. *Biochemistry* **25:** 2248.

Swenson, K.I., K.M. Farrell, and J.V. Ruderman. 1986. The clam embryo protein cyclin A induces entry into M phase and the resumption of meiosis in *Xenopus* oocytes. *Cell* **47:** 861.

Sundin, O. and V. Varshavsky. 1981. Arrest of segregation leads to accumulation in highly intewined catenated dimers: Dissection of the final stages of SV40 DNA replication. *Cell* **25:** 659.

Uemura, T. and M. Yanagida. 1986. Mitotic spindle pulls but fails to separate chromosomes in type II DNA topoisomerase mutants: Uncoordinated mitosis. *EMBO J.* **5:** 1003.

Uemura, T., H. Ohkura, Y. Adachi, K. Morino, K. Shiozaki, and M. Yanagida. 1987. DNA topoisomerase II is required for condensation and separation of mitotic chromosomes in *S. pombe*. *Cell* **50:** 917.

Phosphorylation State of Microtubule-organizing Centers: Regulation of Activity during Mitosis

D.D. Vandre

Department of Biological Sciences, Southern Methodist University
Dallas, Texas 75275-0376

A central question facing cell biologists is to understand at the molecular level the mechanisms involved in the regulation of mitotic processes such as chromosome condensation, nuclear envelope breakdown, and mitotic spindle assembly, function, and subsequent disassembly. The posttranslational modification of proteins by phosphorylation/dephosphorylation reactions is an established mechanism for regulating enzymatic function and other cellular processes. Increasingly, evidence has accumulated indicating the possible role of protein phosphorylation and dephosphorylation in the regulation of mitotic events. Maturation of oocytes has been shown to be induced by the activity of maturation-promoting factor, and partially purified preparations of this factor have an associated kinase activity. Similar maturation activity is also present in extracts prepared from mitotic somatic cells. Furthermore, mitotic activation has been shown to be sensitive to various inhibitors of kinase and phosphatase activities in the oocyte system. The interphase-mitosis transition therefore appears to be regulated, in part, by a common class of mitosis-promoting factors that are characteristically associated with kinase activity (for review, see Schlegel et al. 1987). Phosphorylation is also implicated in the regulation of specific mitotic processes. Nuclear lamina protein phosphorylation is directly involved in nuclear envelope breakdown (Suprynowicz and Gerace 1986), whereas the phosphorylation of histones and vimentin appear to be involved in chromosome condensation and the rearrangement of intermediate filaments, respectively.

The microtubule network is also reorganized during mitosis, from the generalized cytoplasmic microtubule array that is present in interphase cells to the highly organized spindle of mitotic cells. This reorganization may be controlled by alterations in microtubule-associated proteins (MAPs) that affect the assembly and/or stability of individual microtubules and by alterations in components of the microtubule-organizing centers (MTOCs) that may affect either the microtubule-nucleating capacity of the MTOC or its capacity to capture or anchor microtubule ends. These putative alterations also appear to be linked to protein phosphorylation. The phosphorylation state of MAPs has been clearly demonstrated to influence the assembly and stability of microtubules in vitro. Recently, a 215-kD protein, XMAP, has been identified in extracts of *Xenopus* eggs, which promotes the assembly of microtubules, and the activity of the XMAP appears to be regulated through the cell cycle by its phosphorylation state (Gard and Kirschner 1987). Using an immunological approach employing the monoclonal antibody MPM-2, which is reactive with a subset of mitotic phosphoproteins, we have localized antigenically reactive phosphoproteins to structural components of mitotic MTOCs, namely, centrosomes, kinetochores, and mid-bodies (Vandre et al. 1984). Both the spatial distribution of the MPM-reactive phosphoproteins and their temporal appearance upon entry into mitosis correlate with the increased microtubule-nucleating capacity of the centrosome observed during mitosis. There is also a corresponding relationship with the dephosphorylation of these spindle phosphoproteins and the decrease in centrosomal nucleating activity during exit from mitosis (Vandre and Borisy 1985). Thus, the phosphorylation/dephosphorylation of MPM-reactive spindle components may reflect the activity of mitosis-specific protein kinases and phosphoprotein phosphatases. Additional evidence is presented here indicating that the phosphorylation state of centrosomal components is functionally involved in the regulation of centrosomal microtubule nucleation, in both the entry and the exit from mitosis.

RESULTS

Immunofluorescence staining of the pig kidney epithelial cell line LLC-PK with the MPM-2 antibody revealed a specific localization to mitotic MTOCs and spindle fibers, following detergent lysis of the cells prior to fixation. In particular, centrosomes were stained intensely in mitotic cells. Quantitation of the fluorescence staining obtained from cells at different stages of mitosis showed a cyclical change in the relative intensity of the staining specific to the centrosome and, by correlation, the amount of MPM-reactive phosphoproteins associated with the centrosome. The level of staining associated with prophase centrosomes increased dramatically from a basal level of interphase staining. The level of centrosomal staining reached a maximum at metaphase and decreased through anaphase, returning to the interphase levels in late telophase cells.

The increase and subsequent decrease in the levels of centrosomal MPM staining through mitosis indicated that there was a phosphorylation of centrosomal components upon entry into mitosis, followed by their dephosphorylation upon exit from mitosis. This correlates with the increase and decrease in centrosomal microtubule-nucleating capacity observed in vitro. However, direct evidence that phosphorylation of specific centrosomal components was functionally involved in the observed increase in microtubule nucleation could not be established from these observations. In an attempt to determine the functional role for phosphorylation in the nucleation of microtubules by the centrosome, an in vitro microtubule-nucleation assay was developed for the LLC-PK cells similar to that used for other cell types (Brinkley et al. 1981). Briefly, LLC-PK cells grown on glass coverslips were treated with nocodazole to depolymerize the existing microtubule network and spindles. Cells were then lysed with detergent and washed extensively to remove any endogenous tubulin. Lysed cells were further incubated in buffer with and without alkaline phosphatase for 30 minutes at room temperature. After washing out the phosphatase, exogenous pig brain tubulin was applied and microtubules were allowed to regrow for various periods of time at 37°C. Microtubule regrowth and centrosome phosphorylation levels were monitored by indirect immunofluorescence.

Both interphase and mitotic centrosomes were capable of nucleating microtubules, but the number of microtubules nucleated from the mitotic centrosomes was significantly greater (Fig. 1a). These results were similar to those found with other in vitro regrowth systems. The intensity of staining by the MPM antibody correlated with the number of microtubules nucleated: low level of staining and few microtubules nucleated from interphase centrosomes, and intense MPM staining with many microtubules nucleated from the mitotic centrosomes (data not presented). When the lysed cells were

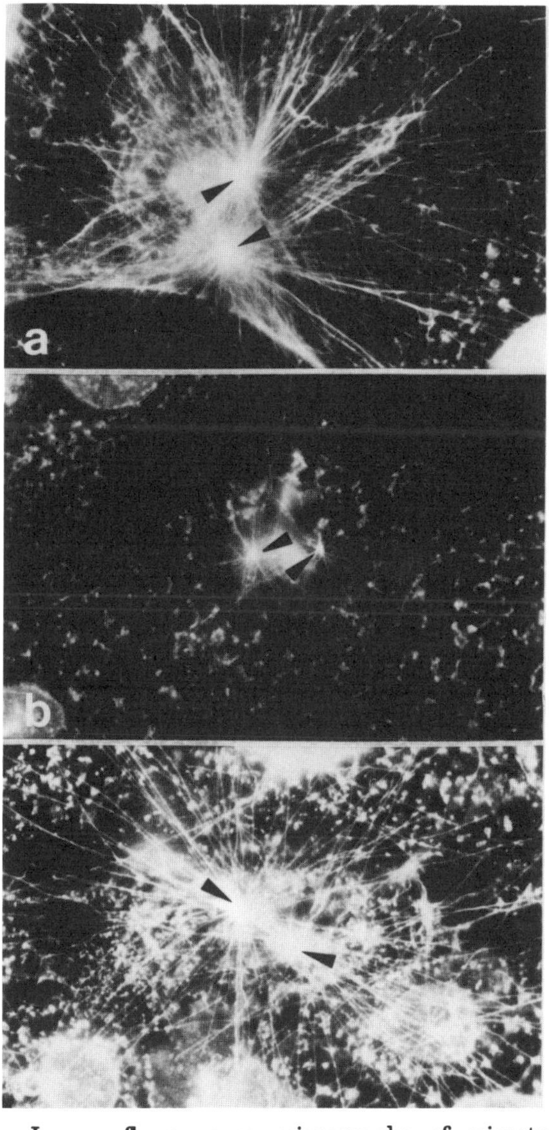

Figure 1 Immunofluorescence micrographs of microtubules nucleated from mitotic centrosomes in detergent-lysed LLC-PK cells and stained with monoclonal antitubulin antibody YL 1/2. (*a*) Regrowth of microtubules at mitotic centrosomes incubated with purified pig brain tubulin. (*b*) Antitubulin staining of cells pretreated with alkaline phosphatase for 30 min prior to the addition of exogenous tubulin. (*c*) Inclusion of β-glycerol phosphate during the incubation with alkaline phosphatase prevented the decrease in the number of nucleated microtubules as a result of phosphatase treatment alone (cf. *b*).

pretreated with alkaline phosphatase, the microtubule-nucleation capacity of the interphase centrosome was abolished (data not presented), and the nucleation capacity of the mitotic centrosome was reduced significantly (Fig. 1b). The level of MPM staining at the centrosome was also substantially reduced in the samples treated with alkaline phosphatase. The reduction in nucleation capacity was demonstrated to be specific for the action of the alkaline phosphatase because inclusion of the phosphatase inhibitor β-glycerol phosphate prevented the loss in microtubule nucleation (Fig. 1c) and the loss of staining by the MPM antibody.

These results strongly suggest that the phosphorylation state of the centrosome modulates its capacity to nucleate microtubules. Increased centrosomal phosphorylation during mitosis appears to be directly involved in the increase in centrosomal microtubule nucleation. Removal of these phosphate groups in vitro decreases the nucleation capacity of the centrosome; therefore, by analogy, the dephosphorylation of the centrosome observed in vivo during anaphase decreases the nucleation capacity of the mitotic centrosome back to interphase levels.

We have recently determined that the phosphatases involved in this process in vivo were not activated until anaphase onset (Vandre and Borisy 1987). Mitotic cells could be blocked in nocodazole for over 18 hours without any loss in the centrosome staining observed with the MPM antibody. Individual cells allowed to progress to within 1 minute of anaphase and then blocked with nocodazole also did not show any loss in the level of MPM staining. However, the centrosomes were dephosphorylated with a halftime of 6.5 minutes when the cells were allowed to start anaphase prior to the nocodazole block. The addition of nocodazole blocked any further chromosome movement and disrupted the spindle, indicating that dephosphorylation was not dependent on either anaphase chromosome motion or spindle integrity. Thus, the phosphorylation state of the centrosome appears to be tightly regulated during the cell cycle and functionally involved in the microtubule nucleation of the centrosome.

DISCUSSION

The biochemical characterization of cellular proteins that are involved in the control of cell proliferation and the identification of the genes that code for these proteins are essential goals for an understanding of the mechanism underlying growth regulation and cell division. Key components of this regulatory system are those proteins involved in modulating the activity of cytoskeletal elements during the cell cycle, because accurate partitioning of chromosomal material requires the proper assembly, functioning, and disassembly of the mitotic apparatus. The phosphorylation state of the centrosome appears to be directly involved in modulating the activity of this MTOC throughout mitosis. The identification and biochemical characterization of the specific centrosomal components phosphorylated during mitosis will be an important step in understanding the molecular basis of centrosomal function.

The phosphorylation/dephosphorylation of MPM-reactive spindle components reflects the activity of mitosis-specific protein kinases and phosphoprotein phosphatases. The establishment of a functional assay for the activity of these enzymes, that of centrosomal microtubule nucleation in vitro, may provide a useful system for the identification and purification of these proteins. For example, cellular extracts from metaphase cells should contain kinases capable of phosphorylating and restoring the microtubule-nucleating capacity of mitotic centrosomes dephosphorylated in vitro by alkaline phosphatase. Conversely, extracts from anaphase or early G_1 cells should contain phosphatases capable of reducing the microtubule-nucleating capacity of mitotic centrosomes. Preliminary experiments suggest that mitotic cell extracts are, in fact, capable of restoring some of the microtubule nucleation of dephosphorylated centrosomes in vitro. Together with the characterization of their target proteins, the analysis of the mitosis-specific kinases and phosphatases will provide necessary information in understanding the process of mitosis.

REFERENCES

Brinkley, B.R., S.M. Cox, D.A. Pepper, L. Wible, S.L. Brenner, and R.L. Pardue. 1981. Tubulin assembly sites and the organization of cytoplasmic microtubules in cultured mammalian cells. *J. Cell. Biol.* **90:** 554.

Gard, D.L. and M.W. Kirschner. 1987. A microtubule-associated protein from *Xenopus* eggs that specifically promotes assembly at the plus end. *J. Cell Biol.* **105:** 2203.

Schlegel, R.A., M.S. Halleck, and P.N. Rao, eds. 1987. *Molecular regulation of nuclear events in mitosis and meiosis.* Academic Press, New York.

Suprynowicz, F.A. and L. Gerace. 1986. A fractionated cell-free system for the analysis of prophase nuclear disassembly. *J. Cell Biol.* **103:** 2073.

Vandre, D.D. and G.G. Borisy. 1985. The interphase-mitosis transformation of the microtubule network in mammalian cells. In *Cell motility: Mechanism and regulation* (ed. H. Ishikawa et al.), p. 389. Univiersity of Tokyo Press, Tokyo.

———. 1987. Anaphase onset and dephosphorylation of mitotic phosphoproteins occur concomitantly. *J. Cell Biol.* **105:** 175a.

Vandre, D.D., F.M. Davis, P.N. Rao, and G.G. Borisy. 1984. Phosphoproteins are components of mitotic microtubule organizing centers. *Proc. Natl. Acad. Sci.* **81:** 4439.

The Fission Yeast Cytoskeleton and Cell-cycle Control

J.S. Hyams, C.E. Alfa, I.M. Hagan, and J. Marks

Department of Biology, University College London
London WC1E 6BT, England

We are investigating the structural rearrangements of the cytoskeletal proteins actin and tubulin, which accompany progress through the fission yeast cell-division cycle. This study aims (1) to determine the role of the cytoskeleton in yeast cell growth and division; (2) to characterize those proteins that associate specifically with actin and tubulin in fission yeast, their relationship to analogous proteins in other cells, and their distribution through the cell cycle; (3) to investigate the relationship between changes in cytoskeletal organization and the functions of known cell-division cycle genes; and (4) to provide new morphological markers for the fission yeast cell-division cycle.

The sausage-shaped cells of *Schizosaccharomyces pombe* grow only by elongation. For the first 0.34 of the cell cycle, growth is confined almost exclusively to the old end (the end that existed before septation). At a point in G_2 termed NETO (Mitchison and Nurse 1985), growth at the new end—the end of the cell recently created by the action of the septum and which hitherto made a negligible contribution to cell extension—is accelerated. Bipolar growth, albeit still biased toward the old end, is maintained until 0.75 of the cell cycle, at which point growth ceases and mitosis is initiated.

Fluorescence microscopy using rhodamine-conjugated phalloidin reveals that changes in the cellular distribution of F-actin through the cell-division cycle parallel precisely the changing pattern of cell-wall deposition (Marks and Hyams 1985). Fluorescence is notably in the form of dots, presumably actin-coated, Golgi-derived vesicles transporting cell-wall precursor molecules (K. Tanaka, pers. comm.). Prior to NETO, the dots are confined almost exclusively to the single growing (old) end. Thereafter, bipolar staining is observed, although the old end remains noticeably brighter.

As growth ceases, actin disappears from the cell ends, and concurrently a ring of actin forms at the cell equator, overlying the nucleus. The function of the ring is to position the septum. Through the use of septation mutants such as *cdc11*$^-$, it has been possible to show that the position of the actin ring is determined by the position of the premitotic nucleus (Marks et al. 1986). The association between nucleus and actin ring appears to involve the product of the *ben4*$^+$ gene (Roy and Fantas 1982; Marks et al. 1986). The relationship between *ben4*$^+$ and actin was emphasized by the construction of strains carrying a duplication of the actin gene (wild-type *S. pombe* cells possess a single actin gene [Mertins and Gallwitz 1987]). Introduction of an extra copy of the actin gene into a *ben4*$^-$ strain is lethal, whereas wild-type cells carrying an actin gene duplication are viable, although they have a higher cell plate index, display a low level (15%) of aberrant septa, and tend to form short chains (A. Schroeder-Lorenz et al., unpubl.).

A number of the *S. pombe* septation mutants, such as *cdc3*$^-$, *cdc4*$^-$, and *cdc12*$^-$, show aberrant organization of actin when grown at the restrictive temperature. At mitosis, actin relocates toward the middle of the cell but appears disorganized rather than as a discrete ring. Such cells deposit a correspondingly aberrant and nonfunctional septum (Marks et al. 1987). It seems likely that these genes encode some type of acting-bundling protein.

Immunofluorescence microscopy of *S. pombe*, using antitubulin antibodies, has revealed the presence of between four and eight bundles of microtubules extending between the cell ends (Hagan and Hyams 1988), with each bundle probably consisting of two to three microtubules. At the G_2/M boundary, this cytoplasmic microtubule complex disassembles rapidly to be replaced by an intranuclear spindle. The end of mitosis is signaled by the elongation of the spindle to the ends of the cell and the appearance of two dots of fluorescence at the cell equator. Microtubules are nucleated from these foci, which appear to be the microtubule-organizing centers (MTOCs) for the interphase cytoplasmic array. These MTOCs are not associated with the nucleus; i.e., they are not the spindle pole bodies that organize the microtubules of the mitotic spindle.

Examination of various *cdc*$^-$ strains at the restrictive temperature showed microtubule patterns consistent with their known cell-cycle block points. For instance, *cdc2*$^-$ and *cdc25*$^-$, which arrest in late G_2, become highly elongated and reveal an exaggerated version of the interphase wild-type cytoplasmic microtubule pattern (Hagan and Hyams 1988). *cdc13⁻* also arrests with an interphase microtubule array, despite the fact that it initiates chromosome condensation (Nasmyth and Nurse 1981). Arrested *cdc13⁻* cultures are heterogeneous with respect to their actin distribution (Marks and Hyams 1985).

Western blots of whole-cell extracts of *S. pombe* or of partially purified tubulin preparations obtained by ion-exchange chromatography reveal the presence of three tubulin isotypes, α1, α2, and β, the products of the two α- and single β-tubulin genes (Yanagida 1987). The carboxy-terminal amino acids of these proteins are Glu-Glu-Tyr, -Glu-Glu-Tyr, and -Leu-Glu-Tyr, respectively. Antibodies specific to the tyrosinated form of α-tubulin (Tyr tubulin; Kreis 1987) recognize both α species but have no reaction with β-tubulin, despite the presence of the carboxy-terminal tyrosine. An antibody specific to detyrosinated α-tubulin (Glu tubulin) has no reactivity. Immunofluorescence microscopy using these antibodies shows that all classes of microtubules in *S. pombe* contain solely tyrosinated α-tubulin. (C.E. Alfa and J.S. Hyams, unpubl.).

REFERENCES

Hagan, I. and J.S. Hyams. 1988. The use of cell division cycle mutants to investigate the control of microtubule distribution in the fission yeast *Saccharomyces pombe*. *J. Cell Sci.* **89:** 343.

Kreis, T.E. 1987. Microtubules containing detyrosinated tubulin are less dynamic. *EMBO J.* **6:** 2597.

Marks, J. and J.S. Hyams. 1985. Localization of F-actin through the cell division cycle of *Schizosaccharomyces pombe*. *Eur. J. Cell Biol.* **39:** 27.

Marks, J., I.M. Hagan, and J.S. Hyams. 1986. Growth polarity and cytokinesis in fission yeast: The role of the cytoskeleton. *J. Cell Sci.* (Suppl.) **5:** 229.

———. 1987. Spatial association of F-actin with growth polarity and septation in the fission yeast *Schizosaccharomyces pombe*. In *Spatial organization in eukaryotic microbes* (ed. R.K. Poole and A.P.J. Trinci), vol. 23, p. 119. Special Publications of the Society for General Microbiology, IRL Press, Oxford.

Mertins, P. and D. Gallwitz. 1987. A single intronless actin gene in the fission yeast *Schizosaccharomyces pombe:* Nucleotide sequence and transcripts formed in homologous and heterologous yeast. *Nucleic Acids Res.* **15:** 7369.

Mitchison, J.M. and P. Nurse. 1985. Growth in cell length in the fission yeast *Schizosaccharomyces pombe*. *J. Cell Sci.* **75:** 357.

Nasmyth, K. and P. Nurse. 1981. Cell division cycle mutants altered in DNA replication and mitosis in the fission yeast *Schizosaccharomyces pombe*. *Mol. Gen. Genet.* **182:** 119.

Roy, D. and P.A. Fantes. 1982. Benomyl resistant mutants of

Schizosaccharomyces pombe cold sensitive for mitosis. *Curr. Genet.* **6:** 195.

Toda, T., Y. Adachi, and M. Yanagida. 1984. Identification of the pleiotrophic cell division cycle gene *NDA2* as one of two different α-tubulin genes in *Schizosacharomyces pombe. Cell* **37:** 233.

Yanagida, M. 1987. Yeast tubulin genes. *Microbiol. Sci.* **4:** 115.

Evidence for the Coupling of Sister Chromatid Separation during Mitotic Anaphase and the Decrease of Intracellular cAMP Concentration in *Schizosaccharomyces pombe*

T. Hirano, H. Ohkura, N. Kinoshita, T. Toda, and M. Yanagida

Department of Biophysics, Faculty of Science, Kyoto University
Sakyo-ku, Kyoto 606, Japan

The disjoining of individual eukaryotic chromosomes in mitotic anaphase A appears to be highly concerted so that the error due to nondisjunction or loss can be kept to a minimum. The process differs fundamentally from that in prokaryotes containing a single chromosome. To answer the question of how multiple chromosomes in eukaryotic cells are concertedly disjoined, we employed the fission yeast *Schizosaccharomyces pombe*, in which the mitotic steps analogous to those found in higher eukaryotes take place. We present results that support the hypothesis that the sister chromatid separation during mitotic anaphase is coupled to the transient decrease of intracellular cAMP concentration during *S. pombe* mitosis.

Phenotype of the *nuc2* Mutant and the Nature of the *nuc2*$^+$ Gene Product

The temperature-sensitive mutant *nuc2-663* specifically blocks mitotic spindle elongation at the restrictive temperature (Hirano et al. 1988). Nuclei in arrested cells contain a short uniform spindle (~3 μm long) that runs through the middle of the three condensed chromosomes, which are assembled into a flat, plate-like structure similar to the metaphase plate (Fig. 1). The spindle axis is perpendicular to the plate. The arrested stage in *nuc2* resembles metaphase. In wild-type or mutant cells at permissive temperature, the spindle is approximately 15 μm long in anaphase B when fully extended.

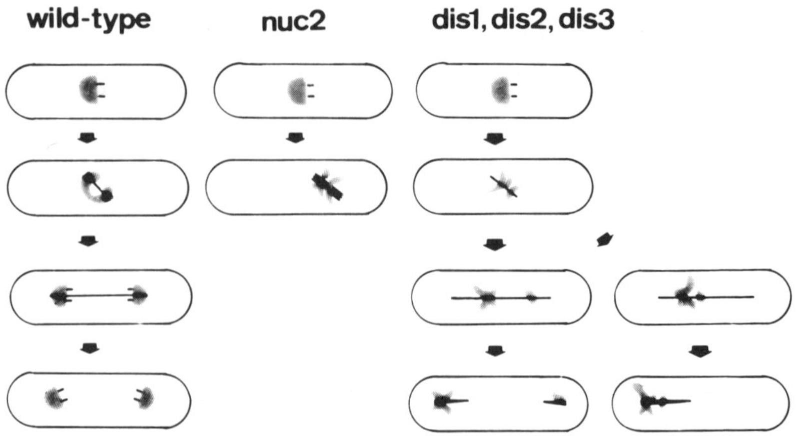

Figure 1 Nuclear division in wild-type, *nuc2*⁺, and *dis*⁺ mutant cells of *S. pombe*. In the *nuc2* cells, chromosomes condense and arrange into a flat, plate-like structure. The spindle is formed but does not elongate. In the *dis* mutants, the spindle forms and elongates, but chromosomes do not separate; they distribute unequally at the ends of the cells at either 2:1 or 3:0.

We have identified the product of the $nuc2^+$ gene in *S. pombe* extracts by immunochemical analyses, using antisera against the $lacZ$-$nuc2^+$ fusion proteins produced in *Escherichia coli* by the cloned $nuc2^+$ gene ligated with the *lacZ* expression plasmid (Hirano et al. 1988). The product of the $nuc2^+$ gene is a minor protein (m.w. 67,000; designated p67) present in the nuclear scaffold fraction, which is insoluble in 2 M NaCl and 25 mM LIS. However, p67 may be the processed form of the direct $nuc2^+$ product, which is soluble and has a higher molecular weight (m.w. 76,000).

Although the $nuc2^+$ gene product located in the nucleus becomes essential in anaphase spindle elongation and chromosome separation, its role at the molecular level is only theorized. Two possibilities are mentioned here (Hirano et al. 1988). First, p67 may interact with microtubules or the spindle apparatus. It may not be required for the initiation of spindle assembly but may become essential in the spindle elongation. It may be involved in the addition of tubulin molecules to the spindle apparatus or in the force generating system for spindle elongation. Second, p67 may interact with a chromosomal domain that is altered during anaphase. It may be a kinetochore protein that plays an essential role in the separation of sister kinetochores. Certain kinetochore proteins are known to behave as nuclear scaffold proteins (Earnshaw et al. 1984).

The *nuc2* Mutation Is Supersensitive to Caffeine

We found that the *nuc2* mutant at permissive temperature is much more sensitive to caffeine than the wild type is, and this supersensitivity could be suppressed by the cloned *nuc2*$^+$ gene (T. Hirano et al., in prep.). Furthermore, the *ts*$^+$ revertants simultaneously suppressed caffeine sensitivity. The *nuc2* cells incubated at permissive temperature in the presence of caffeine showed the phenotype to be similar to that observed at restrictive temperature, whereas the wild-type cells grew normally in the same concentration of the drug at both temperatures. Apparently, the mutant *nuc2* gene product is inactivated directly or indirectly by caffeine. The significance of this finding is discussed below.

Isolation and Characterization of *dis* Mutants

We recently isolated novel classes of cold-sensitive mutants (designated *dis* [*d*efective *i*n *s*ister chromatid separation]) that block chromosome disjoining (Ohkura et al. 1988). Nine strains are mapped in the *dis1* gene, and one each is mapped in the *dis2* and *dis3* genes. The defective phenotype at restrictive temperature (20°C) shown in Figure 1 is similar among the mutants. The chromosomes condense and anomalously move to the cell ends in the absence of disjoining so that they distribute unequally at the two cell ends (2:1 or 3:0). The spindle forms and elongates in the absence of chromosome separation. No defect is found in DNA synthesis. Synchronous culture analyses indicate that the cells can enter into mitosis at the normal timing but become lethal during mitosis. As compared with the wild-type mitosis, differences are found in the spindle formation, the mitotic chromosome structure, the poleward chromosome movement by the spindle elongation, and the telophase spindle degradation. The *dis* mutants lose an artificial minichromosome at exceedingly higher rates at permissive temperature than wild type does.

dis Mutants Are also Supersensitive to Caffeine

We found that all the *dis* mutants isolated are supersensitive to caffeine at permissive temperature (Ohkura et al. 1988). Certain concentrations of caffeine block the growth of the *dis* mutants but not the growth of wild-type. Like the *nuc2* mutant, the phenotype can be reversed by transforming the *dis* mutants with the cloned *dis*⁺ genes, and the same cytological phenotype was obtained by incubation at restrictive temperature or in the presence of caffeine at permissive temperature. *dis* gene expression appears to be affected by caffeine directly or indirectly.

The physiological effect of caffeine may be pleiotropic, but it is a known inhibitor of cAMP phosphodiesterase and increases the intracellular concentration of cAMP. Consistently, we found that the intracellular concentration of cAMP increases in wild-type *S. pombe* by the addition of caffeine (T. Hirano et al., in prep.). Although not conclusive, the *dis* mutants appear to be more sensitive to cAMP than the wild type is (H. Ohkura et al., unpubl.). Furthermore, multicopy plasmids carrying *Saccharomyces cerevisiae* phosphodiesterase (PDE) genes (Sass et al. 1986) complemented *dis* mutations. These results suggest that the rise in intracellular cAMP concentration by caffeine might inactivate the mutant *dis* gene products effectively at permissive temperature. Alternatively, the inactivation of the *dis* gene products might cause the rise of intracellular concentration. We suggest that the dis^+ genes are required for the sister chromatid separation at the time of mitosis and that caffeine might be implicated in the dis^+ gene expression.

Possible Involvement of cAMP in the Sister Chromatid Separation

Although it is well established that cAMP greatly influences the cell-cycle pathway in *S. cerevisiae*, especially at the G_1 phase (Matsumoto et al. 1982), the notion that it is also involved in sister chromatid separation is novel. In *S. pombe*, the expression of certain meiotic genes is affected by cAMP or caffeine (Beach et al. 1985). Our results suggest that the $nuc2^+$ and the three dis^+ gene functions, and concomitantly the sister chromatid, may be related to the intracellular concentration of cAMP directly or indirectly.

We have measured the cAMP concentration in synchronized *S. pombe* wild-type cells. Preliminary results show that the level of cAMP decreases transiently in mitosis (T. Hirano et al., in prep.). Timing for the minimal level of cAMP was coincidental with the maximal frequency of mitotic cells containing separating chromosomes. Further investigation is required to determine the levels of cAMP in the mutant cells.

Although *nuc2* and *dis* mutants are not identical in their

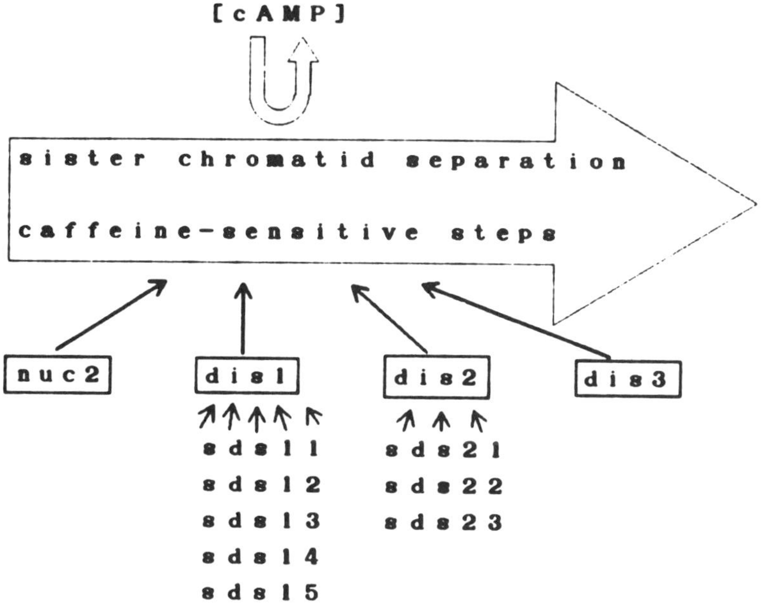

Figure 2 Schematic illustration for the coupling of sister chromatid separation to the temporal decrease of intracellular cAMP concentration during mitosis and the roles of the *nuc2*⁺ and *dis*⁺ genes.

defective phenotype, they commonly block the sister chromatid separation in mitotic anaphase and are also supersensitive to caffeine (Fig. 2). Although indirect, our evidence suggests the coupling of sister chromatid separation to the transient decrease of cAMP in mitosis. In addition to the *dis2*⁺ and *dis3*⁺ genes, we cloned multicopy extragenic suppressor sequences, which complement *dis1* and *dis2* mutations (Ohkura et al. 1988). The predicted polypeptide of a suppressor DNA (*sds22*⁺) is surprisingly very similar to a noncatalytic domain of adenylate cyclase (H. Ohkura et al., in prep.). A complex regulatory system may exist for the execution of the *dis*⁺ gene functions.

ACKNOWLEDGMENTS

This work was supported by grants from the Ministry of Education, Science, and Culture of Japan, and the Mitsubishi Foundation.

REFERENCES

Beach, D.H., L. Rodgers, and J. Gould. 1985. *Ran1*⁺ controls the transition from mitotic division to meiosis in fission yeast. *Curr. Genet.* **10:** 297.

Earnshaw, W.C., N. Halligan, C. Cooke, and N. Rothfield. 1984. The kinetochore is part of the metaphase chromosome scaffold. *J. Cell Biol.* **98:** 352.

Hirano, T., Y. Hiraoka, and M. Yanagida. 1988. A temperature-sensitive mutation of the *Schizosaccharomyces pombe* gene *nuc2⁺* that encodes a nuclear scaffold-like protein blocks spindle elongation in mitotic anaphase. *J. Cell Biol.* **106:** 1171.

Matsumoto, K., I. Uno, Y. Oshima, and T. Ishikawa. 1982. Isolation and characterization of yeast mutants deficient in adenylate cyclase and cAMP-dependent protein kinase. *Proc. Natl. Acad. Sci.* **79:** 2355.

Ohkura, H., Y. Adachi, N. Kinoshita, O. Niwa, T. Toda, and M. Yanagida. 1988. Cold-sensitive and caffeine supersensitive mutants of the *Schizosaccharomyces pombe dis* genes implicated in sister chromatid separation during mitosis. *EMBO J.*. **7:** 1465.

Sass, P., J. Field, J. Nikawa, T. Toda, and M. Wigler. 1986. Cloning and characterization of the high affinity cAMP phosphodiesterase of *S. cerevisiae. Proc. Natl. Acad. Sci.* **83:** 9303.

Regulation of Spindle Pole Duplication in *Saccharomyces*: The Transition from G_1 to S Phase

B. Byers, P. Baum, M.G. Goebl, L. Goetsch, S.L. Johnson, and J. Yochem

Department of Genetics, SK-50, University of Washington
Seattle, Washington 98195

In the yeast cell cycle, a crucial regulatory event characterized by Hartwell et al. (1974) and termed "Start" sets in motion a variety of functions required for cell division. Cloning and molecular analysis have revealed that several of the gene products required at Start exert their functions by serving as effectors and regulators of protein phosphorylation. These include both the class-A Start genes, which mediate cAMP-regulated protein kinase activities (see Matsumoto et al. 1983; Broek et al. 1987), and the class-B genes, of which the archetypical member, *CDC28*, encodes a zinc-dependent protein kinase (see Mendenhall et al. 1987). The Start functions lead to the initiation of several parallel pathways required for bud formation, chromosomal replication, and nuclear division. The molecular mechanisms mediating these subsidiary pathways, which are partially interdependent but have been dissected from one another genetically, remain to be elucidated and are under investigation in our laboratory.

Recent Findings

A notable event in the pathway to nuclear division occurs at Start, when the spindle pole body (SPB) becomes duplicated, apparently by transformation of the attached satellite into a full-fledged daughter SPB. Although this event is spatially and temporally associated with bud emergence and occurs during the initiation of chromosomal DNA replication, the *cdc31* phenotype demonstrates that SPB duplication per se is not directly required for the other events, for mutant cells at elevated temperature specifically fail to form either the satellite or the daughter SPB. Instead, the extant SPB simply becomes enlarged and formation of a monopolar spindle ensues. Sequence analysis of *CDC31* has revealed that its product is similar to those of other calmodulin-like calcium-binding proteins (Baum et al. 1986), and binding of radioactive calcium ions by *trpE-CDC31* fusion proteins confirms the significance of this similarity. These findings suggest that at least this aspect of Start-related regulation involves response to a flux of calcium ions.

The subsequent stage of the spindle formation pathway entails the separation of the two SPBs to form either pole of the mitotic spindle. This function fails to occur in *cdc4* and *cdc34* mutants. The fact that these mutants are also defective in bud maturation (forming numerous abnormally elongate buds at arrest) and in the initiation of chromosomal DNA replication (Hereford and Hartwell 1974) suggests that the roles of these genes are not specific to the SPB cycle but that they globally promote several post-Start functions. Cloning of *CDC4* demonstrated that the carboxy-terminal segment of the derived product contains a repeated pattern of residues that resemble a periodic motif in β-transducin and the β-subunits of other G proteins (Fong et al. 1986; Yochem and Byers 1987). This periodic motif is depicted in Figure 1.

The suggested role for *CDC4* in a transduction system—such as one that might respond to some aspect of Start itself—has induced us to search for related genetic functions. In particular, the phenotypic similarity of *cdc34* suggested that the corresponding wild-type gene might be another component of transducin, and clonal analysis was undertaken. However, the derived sequence of *CDC34* was found not to resemble any components of known G-protein systems but, instead, displayed striking similarity to the yeast *RAD6* product. The significance of this finding became evident when Jentsch et al. (1987) discovered that *RAD6* encodes a ubiquitin-conjugating enzyme that acts on histones H2A and H2B as substrates. A collaborative effort with S. Jentsch, J.P. McGrath, and A. Varshavsky was pursued to determine whether the *CDC34* product might function in a similar manner. We learned, in fact, that the product of this gene, when produced in *Escherichia coli*, is capable of accepting ubiquitin from the ubiquitin-activating enzyme, E1, and then conjugating this ubiquitin to histones H2A and H2B. The reaction appears rather specific, as bacterial proteins present in the same extract are not detectably ubiquitinated. Furthermore, a truncated version of the gene lacking an acid-

```
                                             NH2-MGSFPLAEFPLRDIPV
PYSYRVSGGIASSGSVTALVTAAGTHRNSSTAKTVETEDGEEDIDEYQRKRAAGSGESTP
ERSDFKRVKHDNHKTLHPVNLQNTGAASVDNDGLHNLTDISNDAEKLLMSVDDGSAAPST
LSVNMGVASHNVAAPTTVNAATITGSDVSNNVNSATINNPMEEGALPLSPTASSPGTTTP
LAKTTKTINNNNNIADLIESKDSIISPEYLSDEIFSAINNNLPHAYFKNLLFRLVANMDR
SELSDLGTLIKDNLKRDLITSLPFEISLKIFNYLQFEDIINSLGVSQNWNKIIRKSTSLW
KKLLISENFVSPKGFNSLNLKLSQKYPKLSQQDRLRLSFLENIFILKNWYNPKFVPQRTT

(a) LRGHMTSVITCLQF   EDNYVITGADDKMIRVYDSINKKFLLQ
(b) LSGHDGGVWALKYA   HGGILVSGSTDRTVRVWDIKKGCCTHV
(c) FEGHNSTVRCLDIVEYKNIKYIVTGSRDNTLHVWKLPKESSVPDHG
                                             EEHDYPLVFHTPEENPYFVGV
(d) LRGHMASVRTVSGH   GNIVVSGSYDNTLIVWDVAQMKCLYI
(e) LSGHTDRIYSTIYD  HERKRCISASMDTTIRIWDLENIWNNGECSYA
                                             TNSASPCAKILGAMYT
(f) LQGHTALVGLLRLS   DKFLVSAAADGSIRGWDANDYSRKFSYHHT

NLSAITTFYVSDNILVSGSENQFNIYNLRSGKLVHANILKDADQIWSVNFKGKTLVAAVE
KDGQSFLEILDFSKASKINYVSNPVNSSSSSLESISTSLGLTRTTIIP-COOH


    L.GH...V......(...)...vsgs.D.t.rvWD..... <consensus
```

Figure 1 Amino acid sequence of the *CDC4* product aligned to emphasize the periodic motif within the carboxy-terminal portion, as summarized in the consensus sequence (Yochem and Byers 1987).

rich carboxy-terminal segment remained competent to accept activated ubiquitin from E1 but failed to transfer it to the histones, suggesting that the acidic tail promotes interaction with the basic substrate. Further details are provided in Jentsch et al. (this volume).

Current Directions

The in vitro activity of the *CDC34* product demonstrated in this collaborative work suggests that a similar role may be served in vivo, where the modification might cause a functional change in the chromatin. Ubiquitination of DNA-bound histones might directly affect the potential for replication of the underlying DNA or it could promote expression of specific genes, such as those for the histones themselves, which are known to require *CDC4* activity for their transcription (White et al. 1987). In contrast, proteins other than histones may be the critical substrates for the *CDC34* activity in vivo; the *CDC4* activity itself might fall under the aegis of such ubiquitination.

Preliminary immunochemical investigation of these possibilities has indicated that the *CDC4* product is larger than expected from its gene sequence by about the extent predicted for the addition of a ubiquitin moiety. Furthermore, we have found by immunocytochemistry that the products of both *CDC4* and *CDC34* are directed to the nucleus. Here, the *CDC4* product might serve to modulate the ubiquitin-conjugating activity of *CDC34* toward the histones or other nuclear proteins. Conversely, it remains feasible that the similarity of the *cdc4* and *cdc34* phenotypes reflects a role for *CDC4* in a mechanism distinct from ubiquitination—perhaps one that is more similar to the known signal-transducing mechanisms but is regulated by the *CDC34*-mediated modification of the *CDC4* product. Improved antisera under development should enable us to test these alternatives and establish the specific role of ubiquitination in the transition from G_1 to S phase.

ACKNOWLEDGMENTS

This research was supported by National Institutes of Health (NIH) research grant GM-18541 and NIH training grants AG-00057 and GM-07735.

REFERENCES

Baum, P., C. Furlong, and B. Byers. 1986. Yeast gene required for spindle pole body duplication: Homology of its product with Ca^{++}-binding proteins. *Proc. Natl. Acad. Sci.* **83:** 5512.

Broek, D., T. Toda, T. Michaeli, L. Levin, C. Birchmeier, M. Zoller, S. Powers, and M. Wigler. 1987. The *S. cerevisiae CDC25* gene product regulates the RAS/adenylate cyclase pathway. *Cell* **48:** 789.

Fong, H.K.W., J.B. Hurley, R.S. Hopkins, R. Miake-Lye, M.S. Johnson, R.F. Doolittle, and M.I. Simon. 1986. Repetitive segmental structure of the transducin beta subunit: Homology with the *CDC4* gene and identification of related mRNA's. *Proc. Natl. Acad. Sci.* **83:** 2162.

Hartwell, L.H., J. Culotti, J.R. Pringle, and B.R. Reid. 1974. Genetic control of the cell division cycle in yeast. *Science* **183:** 46.

Hereford, L.M. and L.H. Hartwell. 1974. Sequential gene function in the initiation of *Saccharomyces cerevisiae* DNA synthesis. *J. Mol. Biol.* **84:** 445.

Jentsch, S., J.P. McGrath, and A. Varshavsky. 1987. The yeast DNA repair gene *RAD6* encodes a ubiquitin-conjugating enzyme. *Nature* **329:** 131.

Matsumoto, K., I. Uno, and T. Ishikawa. 1983. Control of cell division in *Saccharomyces cerevisiae.* Mutants defective in adenylate cyclase and cAMP-dependent protein kinase. *Exp. Cell Res.* **146:** 151.

Mendenhall, M.D., C.A. Jones, and S.I. Reed. 1987. Dual regulation of the yeast CDC28-p40 protein kinase complex: Cell cycle, pheromone, and nutrition limitation effects. *Cell* **50:** 927.

White, J.H.M., S.R. Green, D.G. Barker, L.B. Dumas, and L.H. Johnston. 1987. The *CDC8* transcript is cell cycle regulated in

yeast and is expressed coordinately with *CDC9* and *CDC21* at a point preceding histone transcription. *Exp. Cell Res.* **171:** 223.
Yochem, J. and B. Byers. 1987. Structural comparison of the yeast cell division cycle gene *CDC4* and a related pseudogene. *J. Mol. Biol.* **195:** 233.

Regulation of Anaphase Spindle Elongation In Vitro

W.Z. Cande, C. Hogan, and L. Wordeman

Department of Botany, University of California
Berkeley, California 94720

Our understanding of cellular mechanochemistry has been aided by the development of in vitro model systems that mimic key elements of cellular movement in vivo. Isolated diatom spindles undergo anaphase spindle elongation (anaphase B) and have morphological features that make them ideal for analyzing the distribution and movement of spindle components during anaphase (for review, see Cande and McDonald 1985, 1986). In the diatom spindle, the machinery responsible for anaphase A and B is spatially separated. The microtubules associated with chromosome-to-pole movement (anaphase A) are on the periphery of the spindle. The central spindle, responsible for anaphase B, consists of two sets of antiparallel microtubules that are arranged in paracrystalline arrays and overlap in the mid-zone in a highly ordered fashion. Each set of half-spindle microtubules is of relatively uniform length; hence, the zone of microtubule overlap is well defined, and changes in overlap can be readily observed by light microscopy. We have used isolated central spindles prepared from the diatom *Stephanopyxis turris* to describe the mechanochemical events involved in anaphase B and the possible role of phosphorylation of overlap zone-associated proteins in regulating these events.

Spindles Elongate In Vitro because the Two Half-spindles Slide Apart

During spindle reactivation, as visualized on-line with polarization optics, the two half-spindles slide completely apart with a concurrent decrease in the extent of birefringence of the zone of microtubule overlap (Cande and McDonald 1985). Similar structural changes are observed in populations of spindles studied by indirect immunofluorescence and antibodies against tubulin (Cande and McDonald 1986). After ATP addition, a

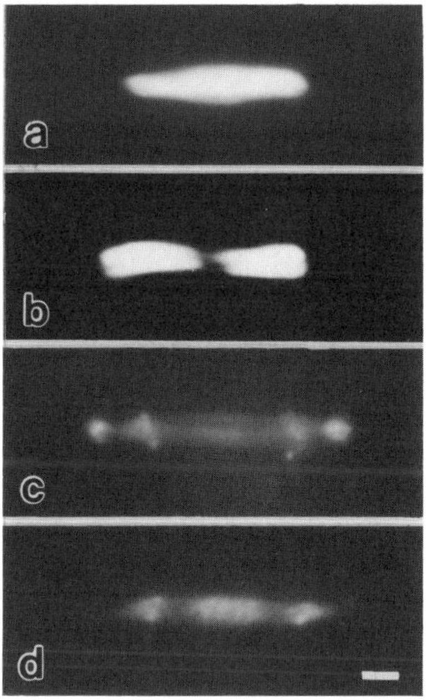

Figure 1 Indirect immunofluorescence of isolated diatom spindles. Antitubulin staining of spindles incubated for 10 min in buffer (*a*) or buffer plus 1 mM ATP (*b*). The spindles in *b* have a noticeable gap between the half-spindles and, on average, are longer than spindles not treated with nucleotide. (*c*) A spindle has been stained with MPM-2 antibody and shows labeling of kinetochores, poles, and the spindle mid-zone. (*d*) Spindles were preincubated in 10 μM γ-S-ATP for 10 min before staining with an antithiophosphate antibody. The mid-zone and kinetochores are labeled. Bar represents 2 μm.

prominent gap develops between the two half-spindles, which is bridged by only a few microtubules (Fig. 1a,b). Populations of spindles that have gaps are longer than control populations by the extent of the original overlap zone (20% of the original spindle length). Spindle reactivation consists of at least three different ATP-dependent events: (1) phosphorylation of spindle-associated proteins by an endogenous spindle-associated kinase, (2) separation of the half-spindles leading to an increase in spindle length, and (3) selective microtubule depolymerization from the overlap zone poleward (Cande and McDonald 1986; Wordeman and Cande 1987). Because almost all spindles undergo these structural changes, it is possible to study the ultrastructure of individual spindles and the biochemistry of bulk spindle extracts with some assurance that the machinery responsible for anaphase B is structurally intact and in a functional state.

Spindle elongation in vitro is not due to tubulin polymerization because spindles elongate in the absence of tubulin, and incorporation of tubulin into isolated spindles does not lead to spindle elongation in the absence of ATP. Nor is spindle elongation due to pulling forces applied to the spindle poles because there are no cytoplasmic structures remaining in vitro attached to the pole complexes that could generate force.

The Role of Tubulin Polymerization during Spindle Elongation

When isolated spindles are incubated in neurotubulin in the presence of ATP, the extent of spindle elongation is no longer limited to the size of the overlap zone, and some spindles elongate by three to four times the original overlap zone length (Masuda and Cande 1987). Biotinylated neurotubulin is incorporated in aster-like arrays at the spindle poles and as broad bands flanking the zone of microtubule overlap. Ultrastructural studies confirm that the newly incorporated tubulin in the mid-zone must be adding on to ends of the original half-spindle microtubules and not forming new microtubules adjacent to the old ones. After ATP addition and the start of spindle elongation, the two bands in the mid-zone merge into one. This process is sensitive to the same inhibitors (vanadate, *N*-ethyl-maleimide) as spindle elongation, and the time course of the rearrangements shows that after newly incorporated tubulin adds on to preexisting microtubule ends, the newly formed polymer slides through the spindle mid-zone (Masuda and Cande 1987). Thus, the mechanism of force generation for anaphase B involves microtubule sliding, and the role of tubulin polymerization is to define the extent of polymer that can slide. Furthermore, for these processes to occur in vitro, the spindle motor must be somehow tethered and confined to the spindle mid-zone, even as microtubules slide through it (L. Wordeman et al., in prep.).

The Diatom Spindle Contains Phosphorylated Proteins

The monoclonal antibody MPM-2 (Davis et al. 1983) recognizes a class of mitosis-specific phosphoproteins, including several spindle-associated proteins (see Vandre et al., this volume). In intact diatom cells, MPM-2 labeling is restricted to microtubule-organizing centers and spindles in mitotic cells (L. Wordeman et al., in prep.). As shown in Figure 1c, MPM-2 labels several discrete structures in isolated diatom spindles, including the pole complex, aggregated kinetochores close to each pole, and the zone of microtubule overlap. These results show that diatom spindles are highly phosphorylated structures in vivo.

In *S. turris,* the highest levels of MPM-2 staining are found in areas putatively associated with microtubule-organizing activity. Similar results have been described in a wide variety of organisms (Vandre et al., this volume). It is possible that the spindle-associated phosphoproteins recognized by the MPM-2 antibody in *S. turris* cells play a role in the dynamic reorganization of microtubules that occurs in these organisms throughout the cell cycle.

Phosphorylation of Overlap Zone Proteins Is Correlated with Spindle Reactivation

Spindles isolated under conditions that permit endogenous phosphatase activity are less capable of undergoing spindle elongation in vitro (Table 1) than spindles in preparations containing phosphatase inhibitors. The ability to elongate is restored to such a preparation by incubating them in low levels of γ-S-ATP, an ATP analog that can be used as a substrate by kinases but does not support spindle elongation (Wordeman and Cande 1987). These experiments demonstrate that preserving the phosphorylated state of spindle proteins is essential for achieving spindle elongation in vitro. An antibody to thiophosphate groups (Gerhart et al. 1985) has allowed us to

Table 1 Effect of Phosphatase Inhibitors on Spindle Reactivation

Treatment (15 min)[a]	Gaps (%)
Ice	47
20°C	13
20°C + 100 μM γ-S-ATP	67
20°C + 40 mM sodium glycerophosphate	31

[a]Mitotic *S. turris* cells were homogenized in a spindle stabilization buffer at pH 7.0, and the crude homogenate was incubated for 15 min in the presence or absence of inhibitors of phosphatase activity. Mitotic nuclei were collected and reactivated in 1 mM ATP for 10 min. Reactivated spindles were scored as those that formed gaps within 10 min. These results indicate that maintaining isolated spindles in a highly phosphorylated state is integral to preserving spindle function in vitro.

visualize directly the distribution of thiophosphorylated proteins in spindles treated with γ-S-ATP. When spindles are incubated in the minimal concentrations of γ-S-ATP required to restore function, the spindle overlap zone and, in some spindles, the kinetochores and part of the pole complex are labeled (Fig. 1d). The major peptide recognized on immunoblots and labeled under these conditions after incubation in [γ-^{35}S]ATP has a molecular weight of 205,000. When spindle microtubules are depolymerized, this protein is released into the supernatant, suggesting that it may be a microtubule-associated protein (MAP) (Wordeman and Cande 1987). During spindle elongation in vitro, the distribution of the mid-zone staining is defined by the size of the overlap zone, and this label remains in the middle of the spindle even as the half-spindles slide apart (L. Wordeman et al., in prep.). Spindles isolated from late anaphase stage cells are usually less intensely labeled with the antithiophosphate antibody than metaphase spindles, suggesting that changes in kinase activity or substrate distribution occur during the mitotic cycle.

Our working hypothesis is that diatom spindle components are substrates for many kinases and that some of these phosphorylations, including the one described above, are involved in regulating spindle function. The spindle-associated kinase activity responsible for thiophosphorylation of overlap zone peptides can utilize γ-S-GTP as a substrate. It does not require physiological levels of calcium and is not blocked by the heat-stable inhibitor of cAMP-dependent kinase (Scott et al. 1985). Washing spindles with high-salt solution removes the kinase but does not affect overall spindle architecture or the distribution of the substrate peptide. This observation should eventually allow us to develop an assay for identifying the spindle-associated kinase.

CONCLUSIONS

We have demonstrated that spindles elongate in vitro because the two half-spindles slide apart, presumably due to forces generated within the zone of microtubule overlap. Tubulin polymerization plays a subsidiary role in this process and is not required to generate the forces that are responsible for spindle elongation. We anticipate that the regulation of spindle function and microtubule dynamics during mitosis is the endpoint in a hierarchy of cell-cycle controls that are regulated by a phosphorylation cascade. The diatom spindle contains many structures with phosphorylated proteins, including proteins in the zone of microtubule overlap. Phosphorylation in vitro of a 205-kD protein located in the overlap zone is correlated with spindle reactivation. This is a direct demonstration that phosphorylation may play a role in the regulation of spindle mechanochemistry.

ACKNOWLEDGMENT

This research was supported by grant GM-23238 from the National Institutes of Health.

REFERENCES

Cande, W.Z. and K.L. McDonald. 1985. *In vitro* reactivation of anaphase spindle elongation using isolated diatom spindles. *Nature* **316:** 168.

———. 1986. Physiological and ultrastructural analysis of elongating mitotic spindles reactivated *in vitro*. *J. Cell Biol.* **103:** 593.

Davis, F.M., T.Y. Tsao, S.K. Fowler, and P.N. Rao. 1983. Monoclonal antibodies to mitotic cells. *Proc. Natl. Acad. Sci.* **80:** 2926.

Gerhart, J., M. Cyert, and M. Kirschner. 1985. M-phase promoting factors from eggs of *Xenopus laevis*. *Cytobios* **43:** 335.

Masuda, H. and W.Z. Cande. 1987. The role of tubulin polymerization during spindle elongation *in vitro*. *Cell* **49:** 193.

Scott, J.D., E.H. Fischer, K. Takio, J.G. Demaille, and E.G. Krebs. 1985. Amino acid sequence of the heat-stable inhibitor of the cAMP-dependent protein kinase from rabbit skeletal muscle. *Proc. Natl. Acad. Sci.* **82:** 5732.

Wordeman, L. and W.Z. Cande. 1987. Reactivation of spindle elongation *in vitro* is correlated with the phosphorylation of a 205kd spindle-associated protein. *Cell* **50:** 535.